PRAISE FOR

THE (HONEST) TRUTH ABOUT DISHONESTY

"Ariely raises the bar for everyone. In the increasingly crowded field of popular cognitive science and behavioral economics, he writes with an unusual combination of verve and sagacity."

—*Washington Post*

"I thought [Ariely's] book was an outstanding encapsulation of the good-hearted and easygoing moral climate of the age."

—David Brooks, *New York Times*

"I was shocked at how prevalent mild cheating was and how much more harmful it can be, cumulatively, compared to outright fraud. This is Dan Ariely's most interesting and most useful book."

—Nassim Nicholas Taleb, author of *The Black Swan*

"Anyone who lies should read this book. And those who claim not to tell lies are liars. So they should read this book too. This is a fascinating, learned, and funny book that will make you a better person."

—A. J. Jacobs, author of *The Year of Living Biblically* and *Drop Dead Healthy*

"Dan Ariely ingeniously and delightfully teases out how people balance truthfulness with cheating to create a reality out of wishful-blindness reality. You'll develop a deeper understanding of your own personal ethics—and those of everybody you know."

—Mehmet Oz, MD, vice-chair and professor of surgery at Columbia University and host of *The Dr. Oz Show*

"The bestselling author's creativity is evident throughout. . . . A lively tour through the impulses that cause many of us to cheat, the book offers especially keen insights into the ways in which we cut corners while still thinking of ourselves as moral people."
—Time.com

"Captivating and astute. . . . In his characteristic spry, cheerful style, Ariely delves deep into the conundrum of human (dis)honesty in the hopes of discovering ways to help us control our behavior and improve our outcomes."
—*Publishers Weekly*

"Ariely writes in a conversational tone one might associate with a popular teacher, providing readers with a working knowledge of what shapes our ethics—or lack thereof."
—*Kirkus Reviews*

"Ariely writes thoughtfully, and his sense of humor is evident throughout the book. A quick and easy read, this is for anyone who wants to learn about the psychological and economic causes of dishonesty."
—*Library Journal*

"Ariely has filled his book with amusing examples of how the cheating psychology plays out in everyday life."
—Bloomberg.com

"Through a remarkable series of experiments, Ariely presents a convincing case that while we all want to view ourselves as honest, we have a strong desire to reap the benefits cheating brings while continuing to view ourselves as 'honest, wonderful people.' . . . Lucid and succinct as always. . . . Required reading for politicians and Wall Street executives."
— *Booklist*

THE (HONEST) TRUTH ABOUT DISHONESTY

ALSO BY DAN ARIELY

The Upside of Irrationality: The Unexpected Benefits of Defying Logic

Predictably Irrational: The Hidden Forces That Shape Our Decisions

THE (HONEST) TRUTH ABOUT DISHONESTY

How We Lie to Everyone—Especially Ourselves

DAN ARIELY

HARPER PERENNIAL

NEW YORK • LONDON • TORONTO • SYDNEY • NEW DELHI • AUCKLAND

HARPER PERENNIAL

A hardcover edition of this book was published in 2012 by HarperCollins Publishers.

HarperCollins books may be purchased for educational, business, or sales promotional use. For information, please e-mail the Special Markets Department at SPsales@harpercollins.com.

FIRST HARPER PERENNIAL EDITION PUBLISHED 2013.

Library of Congress Cataloging-in-Publication Data has been applied for.

ISBN 978-0-06-218361-3 (pbk.)

16 17 DIX/RRD 10 9 8

To my teachers, collaborators, and students,
for making research fun and exciting.

And to all the participants who took part in our experiments over the years—you are the engine of this research, and I am deeply grateful for all your help.

Contents

INTRODUCTION

Why Is Dishonesty So Interesting?

There's one way to find out if a man is honest—ask him. If he says "yes," he is a crook.

—GROUCHO MARX

My interest in cheating was first ignited in 2002, just a few months after the collapse of Enron. I was spending the week at some technology-related conference, and one night over drinks I got to meet John Perry Barlow. I knew John as the erstwhile lyricist for the Grateful Dead, but during our chat I discovered that he had also been working as a consultant for a few companies—including Enron.

In case you weren't paying attention in 2001, the basic story of the fall of the Wall Street darling went something like this: Through a series of creative accounting tricks—helped along by the blind eye of consultants, rating agencies, the company's board, and the now-defunct accounting firm Arthur Andersen, Enron rose to great financial heights only to come crashing down when its actions could no longer be concealed. Stockholders lost their investments, retirement plans evaporated, thousands of employees lost their jobs, and the company went bankrupt.

While I was talking to John, I was especially interested in his description of his own wishful blindness. Even though he consulted for Enron while the company was rapidly spinning out of control, he said he hadn't seen anything sinister going on. In fact, he had fully bought into the worldview that Enron was an innovative leader of the new economy right up until the moment the story was all over the headlines. Even more surprising, he also told me that once the information was out, he could not believe that he failed to see the signs all along. That gave me pause. Before talking to John, I assumed that the Enron disaster had basically been caused by its three sinister C-level architects (Jeffrey Skilling, Kenneth Lay, and Andrew Fastow), who together had planned and executed a large-scale accounting scheme. But here I was sitting with this guy, whom I liked and admired, who had his own story of involvement with Enron, which was one of wishful blindness—not one of deliberate dishonesty.

It was, of course, possible that John and everyone else involved with Enron were deeply corrupt, but I began to think that there may have been a different type of dishonesty at work—one that relates more to wishful blindness and is practiced by people like John, you, and me. I started wondering if the problem of dishonesty goes deeper than just a few bad apples and if this kind of wishful blindness takes place in other companies as well.* I also wondered whether my

---

* The flood of corporate scandals that continued from that point on very clearly answered this question.

friends and I would have behaved similarly if we had been the ones consulting for Enron.

I became fascinated by the subject of cheating and dishonesty. Where does it come from? What is the human capacity for both honesty and dishonesty? And, perhaps most important, is dishonesty largely restricted to a few bad apples, or is it a more widespread problem? I realized that the answer to this last question might dramatically change how we should try to deal with dishonesty: that is, if only a few bad apples are responsible for most of the cheating in the world, we might easily be able to remedy the problem. Human resources departments could screen for cheaters during the hiring process or they could streamline the procedure for getting rid of people who prove to be dishonest over time. But if the problem is not confined to a few outliers, that would mean that anyone could behave dishonestly at work and at home—you and I included. And if we all have the potential to be somewhat criminal, it is crucially important that we first understand how dishonesty operates and then figure out ways to contain and control this aspect of our nature.

WHAT DO WE know about the causes of dishonesty? In rational economics, the prevailing notion of cheating comes from the University of Chicago economist Gary Becker, a Nobel laureate who suggested that people commit crimes based on a rational analysis of each situation. As Tim Harford describes in his book *The Logic of Life*,* the birth of this theory

---

* For the full references to all the materials used in each chapter, and for related readings, see the Bibliography and Additional Readings at the back of the book.

was quite mundane. One day, Becker was running late for a meeting and, thanks to a scarcity of legal parking, decided to park illegally and risk a ticket. Becker contemplated his own thought process in this situation and noted that his decision had been entirely a matter of weighing the conceivable cost—being caught, fined, and possibly towed—against the benefit of getting to the meeting in time. He also noted that in weighing the costs versus the benefits, there was no place for consideration of right or wrong; it was simply about the comparison of possible positive and negative outcomes.

And thus the Simple Model of Rational Crime (SMORC) was born. According to this model, we all think and behave pretty much as Becker did. Like your average mugger, we all seek our own advantage as we make our way through the world. Whether we do this by robbing banks or writing books is inconsequential to our rational calculations of costs and benefits. According to Becker's logic, if we're short on cash and happen to drive by a convenience store, we quickly estimate how much money is in the register, consider the likelihood that we might get caught, and imagine what punishment might be in store for us if we are caught (obviously deducting possible time off for good behavior). On the basis of this cost-benefit calculation, we then decide whether it is worth it to rob the place or not. The essence of Becker's theory is that decisions about honesty, like most other decisions, are based on a cost-benefit analysis.

The SMORC is a very straightforward model of dishonesty, but the question is whether it accurately describes people's behavior in the real world. If it does, society has two clear means for dealing with dishonesty. The first is to increase the probability of being caught (through hiring more police officers and installing more surveillance cameras, for example). The second is to increase the magnitude of punishment for people who get caught (for example, by imposing steeper prison sentences and fines). This, my friends, is the SMORC, with its clear implications for law enforcement, punishment, and dishonesty in general.

But what if the SMORC's rather simple view of dishonesty is inaccurate or incomplete? If that is the case, the standard approaches for overcoming dishonesty are going to be inefficient and insufficient. If the SMORC is an imperfect model of the causes of dishonesty, then we need to first figure out what forces *really* cause people to cheat and then apply this improved understanding to curb dishonesty. That's exactly what this book is about.*

Life in SMORCworld

Before we examine the forces that influence our honesty and dishonesty, let's consider a quick thought experiment. What would our lives be like if we all strictly adhered to the SMORC and considered only the costs and benefits of our actions?

If we lived in a purely SMORC-based world, we would run a cost-benefit analysis on all of our decisions and do what seems to be the most rational thing. We wouldn't make decisions based on emotions or trust, so we would most likely lock our wallets in a drawer when we stepped out of our

---

* Beyond exploring the topic of dishonesty, this book is fundamentally about rationality and irrationality. And although dishonesty is fascinating and important in human endeavors its own right, it is also important to keep in mind that it is but a single component of our interesting and intricate human nature.

office for a minute. We would keep our cash under the mattress or lock it away in a hidden safe. We would be unwilling to ask our neighbors to bring in our mail while we're on vacation, fearing that they would steal our belongings. We would watch our coworkers like hawks. There would be no value in shaking hands as a form of agreement; legal contracts would be necessary for any transaction, which would also mean that we would likely spend a substantial part of our time in legal battles and litigation. We might decide not to have kids because when they grew up, they, too, would try to steal everything we have, and living in our homes would give them plenty of opportunities to do so.

Sure, it is easy to see that people are not saints. We are far from perfect. But if you agree that SMORCworld is not a correct picture of how we think and behave, nor an accurate description of our daily lives, this thought experiment suggests that we don't cheat and steal as much as we would if we were perfectly rational and acted only in our own self-interest.

Calling All Art Enthusiasts

In April 2011, Ira Glass's show, *This American Life,*[1] featured a story about Dan Weiss, a young college student who worked at the John F. Kennedy Center for the Performing Arts in Washington, D.C. His job was to stock inventory for the center's gift shops, where a sales force of three hundred well-intentioned volunteers—mostly retirees who loved theater and music—sold the merchandise to visitors.

The gift shops were run like lemonade stands. There were no cash registers, just cash boxes into which the volunteers deposited cash and from which they made change. The gift shops did a roaring business, selling more than $400,000 worth of merchandise a year. But they had one big problem: of that amount, about $150,000 disappeared each year.

When Dan was promoted to manager, he took on the task of catching the thief. He began to suspect another young employee whose job it was to take the cash to the bank. He contacted the U.S. National Park Service's detective agency, and a detective helped him set up a sting operation. One February night, they set the trap. Dan put marked bills into the cashbox and left. Then he and the detective hid in the nearby bushes, waiting for the suspect. When the suspected staff member eventually left for the night, they pounced on him and found some marked bills in his pocket. Case closed, right?

Not so, as it turned out. The young employee stole only $60 that night, and even after he was fired, money and merchandise still went missing. Dan's next step was to set up an inventory system with price lists and sales records. He told the retirees to write down what was sold and what they received, and—you guessed it—the thievery stopped. The problem was not a single thief but the multitude of elderly, well-meaning, art-loving volunteers who would help themselves to the goods and loose cash lying around.

The moral of this story is anything but uplifting. As Dan put it, "We are going to take things from each other if we have a chance . . . many people need controls around them for them to do the right thing."

THE PRIMARY PURPOSE of this book is to examine the rational cost-benefit forces that are presumed to drive dishonest behavior but (as you will see) often do not, and the irrational forces that we think don't matter but often do. To wit, when a large amount of money goes missing, we usually think it's the work of one coldhearted criminal. But as we saw in the art lovers' story, cheating is not necessarily due to one guy doing a cost-benefit analysis and stealing a lot of money. Instead, it is more often an outcome of many people who quietly justify taking a little bit of cash or a little bit of merchandise over and over. In what follows we will explore the forces that spur us to cheat, and we'll take a closer look at what keeps us honest. We will discuss what makes dishonesty rear its ugly head and how we cheat for our own benefit while maintaining a positive view of ourselves—a facet of our behavior that enables much of our dishonesty.

Once we explore the basic tendencies that underlie dishonesty, we will turn to some experiments that will help us discover the psychological and environmental forces that increase and decrease honesty in our daily lives, including conflicts of interest, counterfeits, pledges, creativity, and simply being tired. We'll explore the social aspects of dishonesty too, including how others influence our understanding of what's right and wrong, and our capacity for cheating when others can benefit from our dishonesty. Ultimately, we will attempt to understand how dishonesty works, how it depends on the structure of our daily environment, and under what conditions we are likely to be more and less dishonest.

In addition to exploring the forces that shape dishonesty, one of the main practical benefits of the behavioral economics approach is that it shows us the internal and environmental influences on our behavior. Once we more clearly understand the forces that really drive us, we discover that we are not helpless in the face of our human follies (dishonesty included), that we can restructure our environment, and that by doing so we can achieve better behaviors and outcomes.

It's my hope that the research I describe in the following chapters will help us understand what causes our own dishonest behavior and point to some interesting ways to curb and limit it.

And now for the journey . . .

CHAPTER 1

Testing the Simple Model of Rational Crime (SMORC)

Let me come right out and say it. They cheat. You cheat. And yes, I also cheat from time to time.

As a college professor, I try to mix things up a bit in order to keep my students interested in the material. To this end, I occasionally invite interesting guest speakers to class, which is also a nice way to reduce the time I spend on preparation. Basically, it's a win-win-win situation for the guest speaker, the class, and, of course, me.

For one of these "get out of teaching free" lectures, I invited a special guest to my behavioral economics class. This clever, well-established man has a fine pedigree: before becoming a legendary business consultant to prominent banks and CEOs, he had earned his juris doctor and, before that, a bachelor's at Princeton. "Over the past few years," I told the class, "our distinguished guest has been helping business elites achieve their dreams!"

With that introduction, the guest took the stage. He was

forthright from the get-go. "Today I am going to help you reach your dreams. Your dreams of MONEY!" he shouted with a thumping, Zumba-trainer voice. "Do you guys want to make some MONEY?"

Everyone nodded and laughed, appreciating his enthusiastic, non-buttoned-down approach.

"Is anybody here rich?" he asked. "I know I am, but you college students aren't. No, you are all poor. But that's going to change through the power of CHEATING! Let's do it!"

He then recited the names of some infamous cheaters, from Genghis Khan through the present, including a dozen CEOs, Alex Rodriguez, Bernie Madoff, Martha Stewart, and more. "You all want to be like them," he exhorted. "You want to have power and money! And all that can be yours through cheating. Pay attention, and I will give you the secret!"

With that inspiring introduction, it was now time for a group exercise. He asked the students to close their eyes and take three deep, cleansing breaths. "Imagine you have cheated and gotten your first ten million dollars," he said. "What will you do with this money? You! In the turquoise shirt!"

"A house," said the student bashfully.

"A HOUSE? We rich people call that a MANSION. You?" he said, pointing to another student.

"A vacation."

"To the private island you own? Perfect! When you make the kind of money that great cheaters make, it changes your life. Is anyone here a foodie?"

A few students raised their hands.

"What about a meal made personally by Jacques Pépin? A wine tasting at Châteauneuf-du-Pape? When you make enough money, you can live large forever. Just ask Donald Trump! Look, we all know that for ten million dollars you would drive over your boyfriend or girlfriend. I am here to tell you that it is okay and to release the handbrake for you!"

By that time most of the students were starting to realize that they were not dealing with a serious role model. But having spent the last ten minutes sharing dreams about all the exciting things they would do with their first $10 million, they were torn between the desire to be rich and the recognition that cheating is morally wrong.

"I can sense your hesitation," the lecturer said. "You must not let your emotions dictate your actions. You must confront your fears through a cost-benefit analysis. What are the pros of getting rich by cheating?" he asked.

"You get rich!" the students responded.

"That's right. And what are the cons?"

"You get caught!"

"Ah," said the lecturer, "There is a CHANCE you will get caught. BUT—here is the secret! Getting caught cheating is not the same as getting punished for cheating. Look at Bernie Ebbers, the ex-CEO of WorldCom. His lawyer whipped out the 'Aw, shucks' defense, saying that Ebbers simply did not know what was going on. Or Jeff Skilling, former CEO of Enron, who famously wrote an e-mail saying, 'Shred the documents, they're onto us.' Skilling later testified that he was just being 'sarcastic'! Now, if these defenses don't work, you can always skip town to a country with no extradition laws!"

Slowly but surely, my guest lecturer—who in real life is a stand-up comedian named Jeff Kreisler and the author of a satirical book called *Get Rich Cheating*—was making a hard case for approaching financial decisions on a purely cost-benefit basis and paying no attention to moral considerations. Listening to Jeff's lecture, the students realized that from a perfectly rational perspective, he was absolutely right. But at the same time they could not help but feel disturbed and repulsed by his endorsement of cheating as the best path to success.

At the end of the class, I asked the students to think about the extent to which their own behavior fit with the SMORC. "How many opportunities to cheat without getting caught do you have in a regular day?" I asked them. "How many of these opportunities do you take? How much more cheating would we see around us if everyone took Jeff's cost-benefit approach?"

Setting Up the Testing Stage

Both Becker's and Jeff's approach to dishonesty are comprised of three basic elements: (1) the benefit that one stands to gain from the crime; (2) the probability of getting caught; and (3) the expected punishment if one is caught. By comparing the first component (the gain) with the last two components (the costs), the rational human being can determine whether committing a particular crime is worth it or not.

Now, it could be that the SMORC is an accurate description of the way people make decisions about honesty and cheating, but the uneasiness experienced by my students (and myself) with the implications of the SMORC suggests that it's worth digging a bit further to figure out what is really going on. (The next few pages will describe in some detail the way we will measure cheating throughout this book, so please pay attention.)

My colleagues Nina Mazar (a professor at the University of Toronto) and On Amir (a professor at the University of California at San Diego) and I decided to take a closer look at how people cheat. We posted announcements all over the MIT campus (where I was a professor at the time), offering students a chance to earn up to $10 for about ten minutes of their time.* At the appointed time, participants entered a room where they sat in chairs with small desks attached (the typical exam-style setup). Next, each participant received a sheet of paper containing a series of twenty different matrices (structured like the example you see on the next page) and were told that their task was to find in each of these matrices two numbers that added up to 10 (we call this the matrix task, and we will refer to it throughout much of this book). We also told them that they had five minutes to solve as many of the twenty matrices as possible and that they would get paid 50 cents per correct answer (an amount that varied depending on the experiment). Once the experimenter said, "Begin!" the participants turned the page over and started solving these simple math problems as quickly as they could.

On the next page is a sample of what the sheet of paper looked like, with one matrix enlarged. How quickly can you find the pair of numbers that adds up to 10?

---

* Readers of *Predictably Irrational* might recognize some of the material presented in this chapter and in chapter 2, "Fun with the Fudge Factor."

Figure 1: Matrix Task

1.69	1.82	2.91
4.67	4.81	3.05
5.82	5.06	4.28
6.36	5.19	4.57

049	0.74	1.17
3.72	2.00	1.22
3.75	5.22	5.67
8.83	8.23	7.70

047	4.58	2.57
3.15	3.82	4.38
4.94	542	5.98
2.95	4.86	7.54

0.17	246	244
6.02	5.60	2.63
6.05	6.21	6.60
8.22	8.19	7.54

046	1.98	2.38
048	1.79	248
0.58	1.69	2.59
1.65	0.98	2.94

0.06	5.07	5.39
1.71	0.03	8.98
2.10	4.96	942
4.53	4.65	9.92

0.85	1.62	1.63
6.06	5.63	1.69
6.25	5.01	1.78
6.36	3.16	1.91

0.15	0.95	1.31
4.98	2.90	2.88
6.66	6.73	7.67
9.75	9.85	8.17

0.63	0.65	1.02
2.64	2.34	2.12
2.89	5.98	8.89
949	9.37	9.33

0.14	0.15	0.32
5.51	5.68	0.52
548	6.15	0.84
5.28	3.31	1.17

0.84	1.54	7.28
442	3.54	7.18
5.54	4.78	5.55
6.99	6.93	6.76

0.77	147	1.69
3.38	3.18	2.28
3.62	3.01	248
3.68	2.93	2.53

0.63	0.74	2.23
8.05	7.68	3.71
8.31	7.06	4.51
845	644	5.29

0.12	0.71	0.74
4.27	3.07	2.27
5.09	5.73	5.82
9.27	7.03	6.79

0.74	1.93	2.76
7.24	5.03	3.14
7.71	6.38	3.8
8.28	9.18	948

0.14	0.67	2.22
5.96	5.58	5.22
7.04	7.59	9.33
9.77	9.50	8.52

0.20	2.54	2.8
1.05	2.39	2.96
144	2.28	3.00
1.73	2.19	3.85

This was how the experiment started for all the participants, but what happened at the end of the five minutes was different depending on the particular condition.

Imagine that you are in the control condition and you are hurrying to solve as many of the twenty matrices as possible. After a minute passes, you've solved one. Two more minutes pass, and you're up to three. Then time is up, and you have four completed matrices. You've earned $2. You walk up to the experimenter's desk and hand her your solutions. After checking your answers, the experimenter smiles approvingly. "Four solved," she says and then counts out your earnings. "That's it," she says, and you're on your way. (The scores in this control condition gave us the actual level of performance on this task.)

Now imagine you are in another setup, called the shredder condition, in which you have the opportunity to cheat. This condition is similar to the control condition, except that after the five minutes are up the experimenter tells you, "Now that you've finished, count the number of correct answers, put your worksheet through the shredder at the back of the room, and then come to the front of the room and tell me how many matrices you solved correctly." If you were in this condition you would dutifully count your answers, shred your worksheet, report your performance, get paid, and be on your way.

If you were a participant in the shredder condition, what would you do? Would you cheat? And if so, by how much?

With the results for both of these conditions, we could compare the performance in the control condition, in which cheating was impossible, to the reported performance in the shredder condition, in which cheating was possible. If the scores were the same, we would conclude that no cheating had occurred. But if we saw that, statistically speaking, people performed "better" in the shredder condition, then we could conclude that our participants overreported their performance (cheated) when they had the opportunity to shred the evidence. And the degree of this group's cheating would be the difference in the number of matrices they claimed to have solved correctly above and beyond the number of matrices participants actually solved correctly in the control condition.

Perhaps somewhat unsurprisingly, we found that given the opportunity, many people did fudge their score. In the control condition, participants solved on average four out of the twenty matrices. Participants in the shredder condition claimed to have solved an average of six—two more than in the control condition. And this overall increase did not result from a few individuals who claimed to solve a lot more matrices, but from lots of people who cheated by just a little bit.

More Money, More Cheating?

With this basic quantification of dishonesty under our belts, Nina, On, and I were ready to investigate what forces motivate people to cheat more and less. The SMORC tells us that people *should* cheat more when they stand a chance of getting more money without being caught or punished. That sounds both simple and intuitively appealing, so we decided to test it next. We set up another version of the matrix experiment, only this time we varied the amount of money the participants would get for solving each matrix correctly. Some participants were promised 25 cents per question; others were promised 50 cents, $1, $2, or $5. At the highest level, we promised some participants a whopping $10 for each correct answer. What do you think happened? Did the amount of cheating increase with the amount of money offered?

Before I divulge the answer, I want to tell you about a related experiment. This time, rather than taking the matrix test themselves, we asked another group of participants to guess how many answers those in the shredder condition would claim to solve correctly at each level of payment. Their predictions were that the claims of correctly solved matrices would increase as the amount of money went up. Essentially, their intuitive theory was the same as the premise of the SMORC. But they were wrong. It turned out that when we looked at the magnitude of cheating, our participants added two questions to their scores on average, regardless of the amount of money they could make per question. In fact, the amount of cheating was slightly *lower* when we promised our participants the highest amount of $10 for each correct answer.

Why wouldn't the level of cheating increase with the amount of money offered? Why was cheating slightly lower at the highest level of payment? This insensitivity to the amount of reward suggests that dishonesty is most likely not an outcome of a cost-benefit analysis. If it were, the increase in the benefit (the amount of money offered) would lead to more cheating. And why was the level of cheating lowest when the payment was greatest? I suspect that when the amount of money that the participants could make per question was $10, it was harder for them to cheat and still feel good about their own sense of integrity (we will come back to this later). At $10 per matrix, we're not talking about cheating on the level of, say, taking a pencil from the office. It's more akin to taking several boxes of pens, a stapler, and a ream of printer paper, which is much more difficult to ignore or rationalize.

To Catch a Thief

Our next experiment looked at what might happen if participants felt that there was a higher probability of getting caught cheating. Basically, we inserted the mental equivalent of a partially operating security camera into the experiment.

We asked one group of participants to shred one half of their worksheet—which meant that if they were dishonest, we might find some evidence of it. We asked a second group to shred the whole work sheet, meaning that they could get off scot-free. Finally, we asked a third group to shred the whole worksheet, leave the testing room, and pay themselves from a sizable bowl of money filled with more than $100 in small bills and coins. In this self-paying condition, participants could not only cheat and get away with it, but they could also help themselves to a lot of extra cash.

Again, we asked a different group to predict how many questions, on average, participants would claim to solve correctly in each condition. Once again, they predicted that the human tendency for dishonesty would follow the SMORC and that participants would claim to solve more matrices as the probability of getting caught decreased.

What did we find? Once again, lots of people cheated, but just by a bit, and the level of cheating was the same across all three conditions (shredding half, shredding all, shredding all and self-paying).

NOW, YOU MIGHT wonder if the participants in our experiments really believed that in our experimental setting, they could cheat and not get caught. To make it clear that this was indeed the case, Racheli Barkan (a professor at Ben-Gurion University of the Negev), Eynav Maharabani (a master's candidate working with Racheli), and I carried out another study where either Eynav or a different research assistant, Tali, proctored the experiment. Eynav and Tali were similar in many ways—but Eynav is noticeably blind, which meant that it was easier to cheat when she was in charge. When it was time to pay themselves from the pile of money that was placed on the table in front of the experimenter, participants could grab as much of the cash as they wanted and Eynav would not be able to see them do so.

So did they cheat Eynav to a greater degree? They still took a bit more money than they deserved, but they cheated just as much when Tali supervised the experiments as they did when Eynav was in charge.

These results suggest that the probability of getting caught doesn't have a substantial influence on the amount of cheating. Of course, I am not arguing that people are entirely uninfluenced by the likelihood of being caught—after all, no one is going to steal a car when a policeman is standing nearby—but the results show that getting caught does not have as great an influence as we tend to expect, and it certainly did not play a role in our experiments.

YOU MIGHT BE wondering whether the participants in our experiments were using the following logic: "If I cheat by only a few questions, no one will suspect me. But if I cheat by more than a small amount, it may raise suspicion and someone might question me about it."

We tested this idea in our next experiment. This time, we told half of the participants that the average student in this experiment solves about four matrices (which was true). We told the other half that the average student solves about eight matrices. Why did we do this? Because if the level of cheating is based on the desire to avoid standing out, then our participants would cheat in both conditions by a few matrices beyond what they believed was the average performance (meaning that they would claim to solve around six matrices when they thought the average was four and about ten matrices when they thought the average was eight).

So how did our participants behave when they expected others to solve more matrices? They were not influenced even to a small degree by this knowledge. They cheated by about two extra answers (they solved four and reported that they had solved six) regardless of whether they thought that others solved on average four or eight matrices.

This result suggests that cheating is not driven by concerns about standing out. Rather, it shows that our sense of our own morality is connected to the amount of cheating we feel comfortable with. Essentially, we cheat up to the level that allows us to retain our self-image as reasonably honest individuals.

Into the Wild

Armed with this initial evidence against the SMORC, Racheli and I decided to get out of the lab and venture into a more natural setting. We wanted to examine common situations that one might encounter on any given day. And we wanted to test "real people" and not just students (though I have discovered that students don't like to be told that they are not real people). Another component missing from our experimental paradigm up to that point was the opportunity for people to behave in positive and benevolent ways. In our lab experiments, the best our participants could do was not cheat. But in many real-life situations, people can exhibit behaviors that are not only neutral but are also charitable and generous. With this added nuance in mind, we looked for situations that would let us test both the negative and the positive sides of human nature.

IMAGINE A LARGE farmer's market spanning the length of a street. The market is located in the heart of Be'er Sheva, a town in southern Israel. It's a hot day, and hundreds of merchants have set out their wares in front of the stores that line both sides of the street. You can smell fresh herbs and sour pickles, freshly baked bread and ripe strawberries, and your eyes wander over plates of olives and cheese. The sound of merchants shouting praises of their goods surrounds you: "*Rak ha yom!*" (only today), "*Matok!*" (sweet), "*Bezol!*" (cheap).

Eynav and Tali entered the market and headed in different directions, Eynav using a white cane to navigate the market. Each of them approached a few vegetable vendors and asked each of the sellers to pick out two kilos (about 4.5 pounds) of tomatoes for them while they went on another errand. Once they made their request, they left for about ten minutes, returned to pick up their tomatoes, paid, and left. From there they took the tomatoes to another vendor at the far end of the market who had agreed to judge the quality of the tomatoes from each seller. By comparing the quality of the tomatoes that were sold to Eynav and to Tali, we could figure out who got better produce and who got worse.

Did Eynav get a raw deal? Keep in mind that from a purely rational perspective, it would have made sense for the seller to choose his worst-looking tomatoes for her. After all, she could not possibly benefit from their aesthetic quality. A traditional economist from, say, the University of Chicago might even argue that in an effort to maximize the social welfare of everyone involved (the seller, Eynav, and the other consumers), the seller should have sold her the worst-looking tomatoes, keeping the pretty ones for people who could also enjoy that aspect of the tomatoes. As it turned out, the visual quality of the tomatoes chosen for Eynav was not worse and, in fact, was superior to those chosen for Tali. The sellers went out of their way, and at some cost to their business, to choose higher-quality produce for a blind customer.

WITH THOSE OPTIMISTIC results, we next turned to another profession that is often regarded with great suspicion: cab drivers. In the taxi world, there is a popular stunt called "long hauling," which is the official term for taking passengers who don't know their way around to their destination via a lengthy detour, sometimes adding substantially to the fare. For example, a study of cab drivers in Las Vegas found that some cabbies drive from McCarran International Airport to the Strip by going through a tunnel to Interstate 215, which can mount to a fare of $92 for what should be a two-mile journey.[1]

Given the reputation that cabbies have, one has to wonder whether they cheat in general and whether they would be more likely to cheat those who cannot detect their cheating. In our next experiment we asked Eynav and Tali to take a cab back and forth between the train station and Ben-Gurion University of the Negev twenty times. The way the cabs on this particular route work is as follows: if you have the driver activate the meter, the fare is around 25 NIS (about $7). However, there is a customary flat rate of 20 NIS (about $5.50) if the meter is not activated. In our setup, both Eynav and Tali always asked to have the meter activated. Sometimes drivers would tell the "amateur" passengers that it would be cheaper not to activate the meter; regardless, both of them always insisted on having the meter activated. At the end of the ride, Eynav and Tali asked the cab driver how much they owed them, paid, left the cab, and waited a few minutes before taking another cab back to the place they had just left.

Looking at the charges, we found that Eynav paid less than Tali, despite the fact that they both insisted on paying by the meter. How could this be? One possibility was that the drivers had taken Eynav on the shortest and cheapest route and had taken Tali for a longer ride. If that were the case, it would mean that the drivers had not cheated Eynav but that they had cheated Tali to some degree. But Eynav had a different account of the results. "I heard the cab drivers activate the meter when I asked them to," she told us, "but later, before we reached our final destination, I heard many of them turn the meter off so that the fare would come out close to twenty NIS." "That certainly never happened to me," Tali said. "They never turned off the meter, and I always ended up paying around twenty-five NIS."

There are two important aspects to these results. First, it's clear that the cab drivers did not perform a cost-benefit analysis in order to optimize their earnings. If they had, they would have cheated Eynav more by telling her that the meter reading was higher than it really was or by driving her around the city for a bit. Second, the cab drivers did better than simply not cheat; they took Eynav's interest into account and sacrificed some of their own income for her benefit.

Making Fudge

Clearly there's a lot more going on here than Becker and standard economics would have us believe. For starters, the finding that the level of dishonesty is not influenced to a large degree (to any degree in our experiments) by the amount of money we stand to gain from being dishonest suggests that dishonesty is not an outcome of simply considering the costs and benefits of dishonesty. Moreover, the results showing that the level of dishonesty is unaltered by changes in the probability of being caught makes it even less likely that dishonesty is rooted in a cost-benefit analysis. Finally, the fact that many people cheat just a little when given the opportunity to do so suggests that the forces that govern dishonesty are much more complex (and more interesting) than predicted by the SMORC.

What is going on here? I'd like to propose a theory that we will spend much of this book examining. In a nutshell, the central thesis is that our behavior is driven by two opposing motivations. On one hand, we want to view ourselves as honest, honorable people. We want to be able to look at ourselves in the mirror and feel good about ourselves (psychologists call this ego motivation). On the other hand, we want to benefit from cheating and get as much money as possible (this is the standard financial motivation). Clearly these two motivations are in conflict. How can we secure the benefits of cheating and at the same time still view ourselves as honest, wonderful people?

This is where our amazing cognitive flexibility comes into play. Thanks to this human skill, as long as we cheat by only a little bit, we can benefit from cheating and still view ourselves as marvelous human beings. This balancing act is the process of rationalization, and it is the basis of what we'll call the "fudge factor theory."

To give you a better understanding of the fudge factor theory, think of the last time you calculated your tax return. How did you make peace with the ambiguous and unclear decisions you had to make? Would it be legitimate to write off a portion of your car repair as a business expense? If so, what amount would you feel comfortable with? And what if you had a second car? I'm not talking about justifying our decisions to the Internal Revenue Service (IRS); I'm talking about the way we are able to justify our exaggerated level of tax deductions to ourselves.

Or let's say you go out to a restaurant with friends and they ask you to explain a work project you've been spending a lot of time on lately. Having done that, is the dinner now an acceptable business expense? Probably not. But what if the meal occurred during a business trip or if you were hoping that one of your dinner companions would become a client in the near future? If you have ever made allowances of this sort, you too have been playing with the flexible boundaries of your ethics. In short, I believe that all of us continuously try to identify the line where we can benefit from dishonesty without damaging our own self-image. As Oscar Wilde once wrote, "Morality, like art, means drawing a line somewhere." The question is: where is the line?

I THINK JEROME K. JEROME got it right in his 1889 novel, *Three Men in a Boat (to Say Nothing of the Dog)*, in which he tells a story about one of the most famously lied-about topics on earth: fishing. Here's what he wrote:

> I knew a young man once, he was a most conscientious fellow and, when he took to fly-fishing, he determined never to exaggerate his hauls by more than twenty-five per cent.
>
> "When I have caught forty fish," said he, "then I will tell people that I have caught fifty, and so on. But I will not lie any more than that, because it is sinful to lie."

Although most people haven't consciously figured out (much less announced) their acceptable rate of lying like this young man, this overall approach seems to be quite accurate; each of us has a limit to how much we can cheat before it becomes absolutely "sinful."

Trying to figure out the inner workings of the fudge factor—the delicate balance between the contradictory desires to maintain a positive self-image and to benefit from cheating—is what we are going to turn our attention to next.

CHAPTER 2

Fun with the Fudge Factor

Here's a little joke for you:

Eight-year-old Jimmy comes home from school with a note from his teacher that says, "Jimmy stole a pencil from the student sitting next to him." Jimmy's father is furious. He goes to great lengths to lecture Jimmy and let him know how upset and disappointed he is, and he grounds the boy for two weeks. "And just wait until your mother comes home!" he tells the boy ominously. Finally he concludes, "Anyway, Jimmy, if you needed a pencil, why didn't you just say something? Why didn't you simply ask? You know very well that I can bring you dozens of pencils from work."

If we smirk at this joke, it's because we recognize the complexity of human dishonesty that is inherent to all of us. We realize that a boy stealing a pencil from a classmate is definitely grounds for punishment, but we are willing to take many pencils from work without a second thought.

To Nina, On, and me, this little joke suggested the possibility that certain types of activities can more easily loosen our moral standards. Perhaps, we thought, if we increased the psychological distance between a dishonest act and its consequences, the fudge factor would increase and our participants would cheat more. Of course, encouraging people to cheat more is not something we want to promote in general. But for the purpose of studying and understanding cheating, we wanted to see what kinds of situations and interventions might further loosen people's moral standards.

To test this idea, we first tried a university version of the pencil joke: One day, I sneaked into an MIT dorm and seeded many communal refrigerators with one of two tempting baits. In half of the refrigerators, I placed six-packs of Coca-Cola; in the others, I slipped in a paper plate with six $1 bills on it. I went back from time to time to visit the refrigerators and see how my Cokes and money were doing—measuring what, in scientific terms, we call the half-life of Coke and money.

As anyone who has been to a dorm can probably guess, within seventy-two hours all the Cokes were gone, but what was particularly interesting was that no one touched the bills. Now, the students could have taken a dollar bill, walked over to the nearby vending machine and gotten a Coke and change, but no one did.

I must admit that this is not a great scientific experiment, since students often see cans of Coke in their fridge, whereas discovering a plate with a few dollar bills on it is rather unusual. But this little experiment suggests that we human beings are ready and willing to steal something that does not explicitly reference monetary value—that is, something that lacks the face of a dead president. However, we shy away from directly stealing money to an extent that would make even the most pious Sunday school teacher proud. Similarly, we might take some paper from work to use in our home printer, but it would be highly unlikely that we would ever take $3.50 from the petty-cash box, even if we turned right around and used the money to buy paper for our home printer.

To look at the distance between money and its influence on dishonesty in a more controlled way, we set up another version of the matrix experiment, this time including a condition where cheating was one step removed from money. As in our previous experiments, participants in the shredder condition had the opportunity to cheat by shredding their worksheets and lying about the number of matrices they'd solved correctly. When the participants finished the task, they shredded their worksheet, approached the experimenter, and said, "I solved X* matrices, please give me X dollars."

The innovation in this experiment was the "token" condition. The token condition was similar to the shredder condition, except that the participants were paid in plastic chips instead of dollars. In the token condition, once participants finished shredding their worksheets, they approached the experimenter and said, "I solved X matrices, please give me X tokens." Once they received their chips, they walked twelve feet to a nearby table, where they handed in their tokens and received cold, hard cash.

As it turned out, those who lied for tokens that a few seconds later became money cheated by about twice as much as

* X stands for the number of questions that the participants claimed to have solved correctly.

those who were lying directly for money. I have to confess that, although I had suspected that participants in the token condition would cheat more, I was surprised by the increase in cheating that came with being one small step removed from money. As it turns out, people are more apt to be dishonest in the presence of nonmonetary objects—such as pencils and tokens—than actual money.

From all the research I have done over the years, the idea that worries me the most is that the more cashless our society becomes, the more our moral compass slips. If being just one step removed from money can increase cheating to such a degree, just imagine what can happen as we become an increasingly cashless society. Could it be that stealing a credit card number is much less difficult from a moral perspective than stealing cash from someone's wallet? Of course, digital money (such as a debit or credit card) has many advantages, but it might also separate us from the reality of our actions to some degree. If being one step removed from money liberates people from their moral shackles, what will happen as more and more banking is done online? What will happen to our personal and social morality as financial products become more obscure and less recognizably related to money (think, for example, about stock options, derivatives, and credit default swaps)?

Some Companies Already Know This!

As scientists, we took great care to carefully document, measure, and examine the influence of being one step removed from money. But I suspect that some companies intuitively understand this principle and use it to their advantage. Consider, for example, this letter that I received from a young consultant:

Dear Dr. Ariely,

I graduated a few years ago with a BA degree in Economics from a prestigious college and have been working at an economic consulting firm, which provides services to law firms.

The reason I decided to contact you is that I have been observing and participating in a very well documented phenomenon of overstating billable hours by economic consultants. To avoid sugar coating it, let's call it cheating. From the most senior people all the way to the lowest analyst, the incentive structure for consultants encourages cheating: no one checks to see how much we bill for a given task; there are no clear guidelines as to what is acceptable; and if we have the lowest billability among fellow analysts, we are the most likely to get axed. These factors create the perfect environment for rampant cheating.

The lawyers themselves get a hefty cut of every hour we bill, so they don't mind if we take longer to finish a project. While lawyers do have some incentive to keep costs down to avoid enraging clients, many of the analyses we perform are very difficult to evaluate. Lawyers know this and seem to use it to their advantage. In effect, we are cheating on their behalf; we get to keep our jobs and they get to keep an additional profit.

Here are some specific examples of how cheating is carried out in my company:

- *A deadline was fast approaching and we were working extremely long hours. Budget didn't seem to be an issue and when I asked how much of my day I should bill, my boss (a midlevel project manager) told me to take the total amount of time I was in the office and subtract two hours, one for lunch and one for dinner. I said that I had taken a number of other breaks while the server was running my programs and she said I could count that as a mental health break that would promote higher productivity later.*

- *A good friend of mine in the office adamantly refused to overbill and consequently had an overall billing rate that was about 20 percent lower than the average. I admire his honesty, but when it was time to lay people off, he was the first to go. What kind of message does that send to the rest of us?*

- *One person bills every hour he is monitoring his email for a project, whether or not he receives any work to do. He is "on-call," he says.*

- *Another guy often works from home and seems to bill a lot, but when he is in the office he never seems to have any work to do.*

These kinds of examples go on and on. There is no doubt that I am complicit in this behavior, but seeing it more clearly makes me want to fix the problems. Do you have any advice? What would you do in my situation?

Sincerely yours,

Jonah

Unfortunately, the problems Jonah noted are commonplace, and they are a direct outcome of the way we think about our own morality. Here is another way to think about this issue: One morning I discovered that someone had broken the window of my car and stolen my portable GPS system. Certainly, I was very annoyed, but in terms of its economic impact on my financial future, this crime had a very small effect. On the other hand, think about how much my lawyers, stockbrokers, mutual fund managers, insurance agents, and others probably take from me (and all of us) over the years by slightly overcharging, adding hidden fees, and so on. Each of these actions by itself is probably not very financially significant, but together they add up to much more than a few navigation devices. At the same time, I suspect that unlike the person who took my GPS, those white-collar transgressors think of themselves as highly moral people because their actions are relatively small and, most important, several steps removed from my pocket.

The good news is that once we understand how our dishonesty increases when we are one or more steps removed from money, we can try to clarify and emphasize the links between our actions and the people they can affect. At the same time, we can try to shorten the distance between our actions and the money in question. By taking such steps, we can become more cognizant of the consequences of our actions and, with that awareness, increase our honesty.

LESSONS FROM LOCKSMITHS

Not too long ago, one of my students named Peter told me a story that captures our misguided efforts to decrease dishonesty rather nicely.

One day, Peter locked himself out of his house, so he called around to find a locksmith. It took him a while to find one who was certified by the city to unlock doors. The locksmith finally pulled up in his truck and picked the lock in about a minute.

"I was amazed at how quickly and easily this guy was able to open the door," Peter told me. Then he passed on a little lesson in morality he learned from the locksmith that day.

In response to Peter's amazement, the locksmith told Peter that locks are on doors only to keep honest people honest. "One percent of people will always be honest and never steal," the locksmith said. "Another one percent will always be dishonest and always try to pick your lock and steal your television. And the rest will be honest as long as the conditions are right—but if they are tempted enough, they'll be dishonest too. Locks won't protect you from the thieves, who can get in your house if they really want to. They will only protect you from the mostly honest people who might be tempted to try your door if it had no lock."

After reflecting on these observations, I came away thinking that the locksmith was probably right. It's not that 98 percent of people are immoral or will cheat any time the opportunity arises; it's more likely that most of us need little reminders to keep ourselves on the right path.

How to Get People to Cheat Less

Now that we had figured out how the fudge factor works and how to expand it, as our next step we wanted to figure out whether we could decrease the fudge factor and get people to cheat less. This idea, too, was spawned by a little joke:

A visibly upset man goes to see his rabbi one day and says, "Rabbi, you won't believe what happened to me! Last week, someone stole my bicycle from synagogue!"

The rabbi is deeply upset by this, but after thinking for a moment, he offers a solution: "Next week come to services, sit in the front row, and when we recite the Ten Commandments, turn around and look at the people behind you. And when we get to 'Thou shalt not steal,' see who can't look you in the eyes and that's your guy." The rabbi is very pleased with his suggestion, and so is the man.

At the next service, the rabbi is very curious to learn whether his advice panned out. He waits for the man by the doors of the synagogue, and asks him, "So, did it work?"

"Like a charm," the man answers. "The moment we got to 'Thou shalt not commit adultery,' I remembered where I left my bike."

What this little joke suggests is that our memory and awareness of moral codes (such as the Ten Commandments) might have an effect on how we view our own behavior.

Inspired by the lesson behind this joke, Nina, On, and I ran an experiment at the University of California, Los Angeles (UCLA). We took a group of 450 participants and split them into two groups. We asked half of them to try to recall the Ten Commandments and then tempted them to cheat on our matrix task. We asked the other half to try to recall ten books they had read in high school before setting them loose on the matrices and the opportunity to cheat. Among the group who recalled the ten books, we saw the typical widespread but moderate cheating. On the other hand, in the group that was asked to recall the Ten Commandments, we observed no cheating whatsoever. And that was despite the fact that no one in the group was able to recall all ten.

This result was very intriguing. It seemed that merely trying to recall moral standards was enough to improve moral behavior. In another attempt to test this effect, we asked a group of self-declared atheists to swear on a Bible and then gave them the opportunity to claim extra earnings on the matrix task. What did the atheists do? They did not stray from the straight-and-narrow path.

STEALING PAPER

A few years ago I received a letter from a woman named Rhonda who attended the University of California at Berkeley. She told me about a problem she'd had in her house and how a little ethical reminder helped her solve it.

She was living near campus with several other people—none of whom knew one another. When the cleaning people came each weekend, they left several rolls of toilet paper in each of the two bathrooms. However, by Monday all the toilet paper would be gone. It was a classic tragedy-of-the-commons situation: because some people hoarded the toilet paper and took more than their fair share, the public resource was destroyed for everyone else.

These experiments with moral reminders suggest that our willingness and tendency to cheat could be diminished if we are given reminders of ethical standards. But although using the Ten Commandments and the Bible as honesty-building mechanisms might be helpful, introducing religious tenets into society on a broader basis as a means to reduce cheating is not very practical (not to mention the fact that doing so would violate the separation of church and state). So we began to think of more general, practical, and secular ways to shrink the fudge factor, which led us to test the honor codes that many universities already use.

To discover whether honor codes work, we asked a group of MIT and Yale students to sign such a code just before giving half of them a chance to cheat on the matrix tasks.

After reading about the Ten Commandments experiment on my blog, Rhonda put a note in one of the bathrooms asking people not to remove toilet paper, as it was a shared commodity. To her great satisfaction, one roll reappeared in a few hours, and another the next day. In the other note-free bathroom, however, there was no toilet paper until the following weekend, when the cleaning people returned.

This little experiment demonstrates how effective small reminders can be in helping us maintain our ethical standards and, in this case, a fully stocked bathroom.

The statement read, "I understand that this experiment falls under the guidelines of the MIT/Yale honor code." The students who were not asked to sign cheated a little bit, but the MIT and Yale students who signed this statement did not cheat at all. And that was despite the fact that neither university has an honor code (somewhat like the effect that swearing on the Bible had on the self-declared atheists).

We found that an honor code worked in universities that don't have an honor code, but what about universities that have a strong honor code? Would their students cheat less all the time? Or would they cheat less only when they signed the honor code? Luckily, at the time I was spending some time at the Institute for Advanced Study at Princeton University, which was a great petri dish in which to test this idea.

Princeton University has a rigorous honor system that's been around since 1893. Incoming freshmen receive a copy of the Honor Code Constitution and a letter from the Honor Committee about the honor system, which they must sign before they can matriculate. They also attend mandatory talks about the importance of the Honor Code during their first week of school. Following the lectures, the incoming Princetonians further discuss the system with their dorm advising group. As if that weren't enough, one of the campus music groups, the Triangle Club, performs its "Honor Code Song" for the incoming class.

For the rest of their time at Princeton, students are repeatedly reminded of the honor code: they sign an honor code at the end of every paper they submit ("This paper represents my own work in accordance with University regulations"). They sign another pledge for every exam, test, or quiz ("I pledge my honor that I have not violated the honor code during this examination"), and they receive biannual reminder e-mails from the Honor Committee.

To see if Princeton's crash course on morality has a long-term effect, I waited two weeks after the freshmen finished their ethics training before tempting them to cheat—giving them the same opportunities as the students at MIT and Yale (which have neither an honor code nor a weeklong course on academic honesty). Were the Princeton students, still relatively fresh from their immersion in the honor code, more honest when they completed the matrix task?

Sadly, they were not. When the Princeton students were asked to sign the honor code, they did not cheat at all (but neither did the MIT or Yale students). However, when they were not asked to sign the honor code, they cheated just as much as their counterparts at MIT and Yale. It seems that the crash course, the propaganda on morality, and the existence of an honor code did not have a lasting influence on the moral fiber of the Princetonians.

These results are both depressing and promising. On the depressing side, it seems that it is very difficult to alter our behavior so that we become more ethical and that a crash course on morality will not suffice. (I suspect that this ineffectiveness also applies to much of the ethics training that takes place in businesses, universities, and business schools.) More generally, the results suggest that it's quite a challenge to create a long-term cultural change when it comes to ethics.

On the positive side, it seems that when we are simply reminded of ethical standards, we behave more honorably. Even better, we discovered that the "sign here" honor code method works both when there is a clear and substantial cost for dishonesty (which, in the case of Princeton, can entail expulsion) and when there is no specific cost (as at MIT and Yale). The good news is that people seem to want to be honest, which suggests that it might be wise to incorporate moral reminders into situations that tempt us to be dishonest.*

ONE PROFESSOR AT Middle Tennessee State University got so fed up with the cheating among his MBA students that he decided to employ a more drastic honor code. Inspired by our Ten Commandments experiment and its effect on honesty, Thomas Tang asked his students to sign an honor code stating that they would not cheat on an exam. The pledge also stated that they "would be sorry for the rest of their lives and go to Hell" if they cheated.

The students, who did not necessarily believe in Hell or agree that they were going there, were outraged. The pledge became very controversial, and, perhaps unsurprisingly, Tang caught a lot of heat for his effort (he eventually had to revert to the old, Hell-free pledge).

Still, I imagine that in its short existence, this extreme version of the honor code had quite an effect on the students. I also think the students' outrage indicates how effective this type of pledge can be. The future businessmen and women must have felt that the stakes were very high, or they would

* One important question about the usage of moral reminders is whether over time people will get used to signing such honor codes, causing such reminders to lose their effectiveness. That is why I think that the right approach is to ask people to write their own version of the honor code—that way it will be difficult to sign without thinking about morality, and more ethical behavior should follow.

not have cared so much. Imagine yourself confronted by such a pledge. How comfortable would you feel signing it? Would signing it influence your behavior? What if you had to sign it just before filling out your expense reports?

RELIGIOUS REMINDERS

The possibility of using religious symbols as a way to increase honesty has not escaped religious scholars. There is a story in the Talmud about a religious man who becomes desperate for sex and goes to a prostitute. His religion wouldn't condone this, of course, but at the time he feels that he has more pressing needs. Once alone with the prostitute, he begins to undress. As he takes off his shirt, he sees his tzitzit, an undergarment with four pieces of knotted fringe. Seeing the tzitzit reminds him of the mitzvoth (religious obligations), and he quickly turns around and leaves the room without violating his religious standards.

Adventures with the IRS

Using honor codes to curb cheating at a university is one thing, but would moral reminders of this type also work for other types of cheating and in nonacademic environments? Could they help prevent cheating on, say, tax-reporting and insurance claims? That is what Lisa Shu (a PhD student at Harvard University), Nina Mazar, Francesca Gino (a professor at Harvard University), Max Bazerman (a professor at Harvard University), and I set out to test.

We started by restructuring our standard matrix experiment to look a bit like tax reporting. After they finished solving and shredding the matrix task, we asked participants to write down the number of questions that they had solved correctly on a form we modeled after the basic IRS 1040EZ tax form. To make it feel even more as if they were working with a real tax form, it was stated clearly on the form that their income would be taxed at a rate of 20 percent. In the first section of the form, the participants were asked to report their "income" (the number of matrices they had solved correctly). Next, the form included a section for travel expenses, where participants could be reimbursed at a rate of 10 cents per minute of travel time (up to two hours, or $12) and for the direct cost of their transportation (up to another $12). This part of the payment was tax exempt (like a business expense). The participants were then asked to add up all the numbers and come up with their final net payment.

There were two conditions in this experiment: Some of the participants filled out the entire form and then signed it at the bottom, as is typically done with official forms. In this condition, the signature acted as verification of the information on the form. In the second condition, participants signed the form first and only then filled it out. That was our "moral reminder" condition.

What did we find? The participants in the sign-at-the-end condition cheated by adding about four extra matrices to their score. And what about those who signed at the top? When the signature acted as a moral reminder, participants claimed only one extra matrix. I am not sure how you feel about "only" one added matrix—after all, it is still cheating—but given that the one difference between these two conditions was the location of the signature line, I see this outcome as a promising way to reduce dishonesty.

Our version of the tax form also allowed us to look at the requests for travel reimbursements. Now, we did not know how much time the participants really spent traveling, but if we assumed that due to randomization, the average amount of travel time was basically the same in both conditions, we could see in which condition participants claimed higher travel expenses. What we saw was that the amount of requests for travel reimbursement followed the same pattern: Those in the signature-at-the-bottom condition claimed travel expenses averaging $9.62, while those in the moral reminder (signature-at-the-top) condition claimed that they had travel expenses averaging $5.27.

ARMED WITH OUR evidence that when people sign their names to some kind of pledge, it puts them into a more honest disposition (at least temporarily), we approached the IRS, thinking that Uncle Sam would be glad to hear of ways to boost tax revenues. The interaction with the IRS went something like this:

ME: By the time taxpayers finish entering all the data onto the form, it is too late. The cheating is done and over with, and no one will say, "Oh, I need to sign this thing, let me go back and give honest answers." You see? If people sign before they enter any data onto the form, they cheat less. What you need is a signature at the top of the form, and this will remind everyone that they are supposed to be telling the truth.

IRS: Yes, that's interesting. But it would be illegal to ask people to sign at the top of the form. The signature needs to verify the accuracy of the information provided.

ME: How about asking people to sign twice? Once at the top and once at the bottom? That way, the top signature will act as a pledge—reminding people of their patriotism, moral fiber, mother, the flag, homemade apple pie—and the signature at the bottom would be for verification.

IRS: Well, that would be confusing.

ME: Have you looked at the tax code or the tax forms recently?

IRS: [*No reaction.*]

ME: How about this? What if the first item on the tax form asked if the taxpayer would like to donate twenty-five dollars to a task force to fight corruption? Regardless of the particular answer, the question will force people to contemplate their standing on honesty and its importance for society! And if the taxpayer donates money to this task force, they not only state an opinion, but they also put some money behind their decision, and now they might be even more likely to follow their own example.

IRS: [*Stony silence.*]

ME: This approach may have another interesting benefit: You could flag the taxpayers who decide not to donate to the task force and audit them!

IRS: Do you really want to talk about audits?*

* As it turned out, I was audited by the IRS a few years later, and it was a long, painful but very interesting experience. I don't think it was related to this meeting.

Despite the reaction from the IRS, we were not entirely discouraged, and continued to look for other opportunities to test our "sign first" idea. We were finally (moderately) successful when we approached a large insurance company. The company confirmed our already substantiated theory that most people cheat, but only by a little bit. They told us that they suspect that very few people cheat flagrantly (committing arson, faking a robbery, and so on) but that many people who undergo a loss of property seem comfortable exaggerating their loss by 10 to 15 percent. A 32-inch television becomes 40 inches, an 18k necklace becomes 22k, and so on.

I went to their headquarters and got to spend the day with the top folks at this company, trying to come up with ways to decrease dishonest reporting on insurance claims. We came up with lots of ideas. For instance, what if people had to declare their losses in highly concrete terms and provide more specific details (where and when they bought the items) in order to allow less moral flexibility? Or if a couple lost their house in a flood, what if they had to agree on what was lost (although as we will see in chapter 8, "Cheating as an Infection," and chapter 9, "Collaborative Cheating," this particular idea might backfire). What if we played religious music when people were on hold? And of course, what if people had to sign at the top of the claim form or even next to each reported item?

As is the way with such large companies, the people I met with took the ideas to their lawyers. We waited six months and then finally heard from the lawyers—who said that they were not willing to let us try any of these approaches.

A few days later, my contact person at the insurance company called me and apologized for not being able to try any of our ideas. He also told me that there was one relatively unimportant automobile insurance form that we could use for an experiment. The form asked people to record their current odometer reading so that the insurance company could calculate how many miles they had driven the previous year. Naturally, people who want their premium to be lower (I can think of many) might be tempted to lie and underreport the actual number of miles they drove.

The insurance company gave us twenty thousand forms, and we used them to test our sign-at-the-top versus the sign-at-the-bottom idea. We kept half of the forms with the "I promise that the information I am providing is true" statement and signature line on the bottom of the page. For the other half, we moved the statement and signature line to the top. In all other respects, the two forms were identical. We mailed the forms to twenty thousand customers and waited a while, and when we got the forms back we were ready to compare the amount of driving reported on the two types of forms. What did we find?

When we estimated the amount of driving that took place over the last year, those who signed the form first appeared to have driven on average 26,100 miles, while those who signed at the end of the form appeared to have driven on average 23,700 miles—a difference of about 2,400 miles. Now, we don't know how much those who signed at the top really drove, so we don't know if they were perfectly honest—but we do know that they cheated to a much lesser degree. It is also interesting to note that this magnitude of decreased cheating (which was about 15 percent of the total amount of driving reported) was similar to the percentage of dishonesty we found in our lab experiments.

TOGETHER, THESE EXPERIMENTAL results suggest that although we commonly think about signatures as ways to verify information (and of course signatures can be very useful in fulfilling this purpose), signatures at the top of forms could also act as a moral prophylactic.

COMPANIES ARE ALWAYS RATIONAL!

Many people believe that although individuals might behave irrationally from time to time, large commercial companies that are run by professionals with boards of directors and investors will always operate rationally. I never bought into this sentiment, and the more I interact with companies, the more I find that they are actually far less rational than individuals (and the more I am convinced that anyone who thinks that companies are rational has never attended a corporate board meeting).

What do you think happened after we demonstrated to the insurance company that we could improve honesty in mileage reporting using their forms? Do you think the company was eager to emend their regular practices? They were not! Or do you think anyone asked (maybe begged) us to experiment with the much more important problem of exaggerated losses on property claims—a problem that they estimate costs the insurance industry $24 billion a year? You guessed it—no one called.

Some Lessons

When I ask people how we might reduce crime in society, they usually suggest putting more police on the streets and applying harsher punishments for offenders. When I ask CEOs of companies what they would do to solve the problem of internal theft, fraud, overclaiming on expense reports, and sabotage (when employees do things to hurt their employer with no concrete benefit to themselves), they usually suggest stricter oversight and tough no-tolerance policies. And when governments try to decrease corruption or create regulations for more honest behavior, they often push for transparency (also known as "sunshine policies") as a cure for society's ills. Of course, there is little evidence that any of these solutions work.

By contrast, the experiments described here show that doing something as simple as recalling moral standards at the time of temptation can work wonders to decrease dishonest behavior and potentially prevent it altogether. This approach works even if those specific moral codes aren't a part of our personal belief system. In fact, it's clear that moral reminders make it relatively easy to get people to be more honest—at least for a short while. If your accountant were to ask you to sign an honor code a moment before filing your taxes or if your insurance agent made you swear that you were telling the whole truth about that water-damaged furniture, chances are that tax evasion and insurance fraud would be less common.*

What are we to make of all this? First, we need to recognize that dishonesty is largely driven by a person's fudge

---

* I suspect that for people who actively dislike the government or insurance companies, the effect would still hold, though it might be mitigated to some degree—something worth testing in the future.

factor and not by the SMORC. The fudge factor suggests that if we want to take a bite out of crime, we need to find a way to change the way in which we are able to rationalize our actions. When our ability to rationalize our selfish desires increases, so does our fudge factor, making us more comfortable with our own misbehavior and cheating. The other side is true as well; when our ability to rationalize our actions is reduced, our fudge factor shrinks, making us less comfortable with misbehaving and cheating. When you consider the range of undesirable behaviors in the world from this standpoint—from banking practices to backdating stock options, from defaulting on loans and mortgages to cheating on taxes—there's a lot more to honesty and dishonesty than rational calculations.

Of course, this means that understanding the mechanisms involved in dishonesty is more complex and that deterring dishonesty is not an easy task—but it also means that uncovering the intricate relationship between honesty and dishonesty will be a more exciting adventure.

CHAPTER 2B

Golf

The income tax has made more liars out of
the American people than golf has.
—WILL ROGERS

There's a scene in the movie *The Legend of Bagger Vance* where Matt Damon's character, Rannulph Junuh, is attempting to get his golf game back, but he makes a critical error and his ball ends up in the woods. After making it back onto the green, he moves a twig that is just adjacent to the ball in order to create a clear path for his shot. As he moves the twig the ball rolls a tiny bit to the side. According to the rules, he has to count it as a stroke. At that point in the match, Junuh had gained enough of a lead that if he ignored the rule, he could win, making a comeback and restoring his former glory. His youthful assistant tearfully begs Junuh to ignore the movement of the ball. "It was an accident," the assistant says, "and it's a stupid rule anyway. Plus, no one would ever know." Junuh turns to him and says stoically, "I will. And so will you."

Even Junuh's opponents suggest that most likely the ball just wobbled and returned to its former position or that the light tricked Junuh into thinking that the ball moved. But Junuh insists that the ball rolled away. The result is an honorably tied game.

That scene was inspired by a real event that occurred during the 1925 U.S. Open. The golfer, Bobby Jones, noticed that his ball moved ever so slightly as he prepared for his shot in the rough. No one saw, no one would ever have known, but he called the stroke on himself and went on to lose the match. When people discovered what he'd done and reporters began to flock to him, Jones famously asked them not to write about the event, saying "You might as well praise me for not robbing banks." This legendary moment of noble honesty is still referred to by those who love the game, and for good reason.

I think this scene—both cinematic and historic—captures the romantic ideal of golf. It's a demonstration of man versus himself, showing both his skill and nobility. Perhaps these characteristics of self-reliance, self-monitoring, and high moral standards are why golf is often used as a metaphor for business ethics (not to mention the fact that so many businesspeople spend so much time on golf courses). Unlike other sports, golf has no referee, umpire, or panel of judges to make sure rules are followed or to make calls in questionable situations. The golfer, much like the businessperson, has to decide for him- or herself what is and is not acceptable. Golfers and businesspeople must choose for themselves what they are willing and not willing to do, since most of the time there is no one else to supervise or check their work. In fact, golf's three underlying rules are, play the ball as it lies, play the course as you find it, and if you cannot do either, do what is fair. But "fair" is a notoriously difficult thing to determine. After all, a lot of people might judge not counting an accidental and inconsequential change in the ball's location after a movement of a twig as "fair." In fact, it might seem pretty unfair to be penalized for an incidental movement of the ball.

DESPITE THE NOBLE heritage that golfers claim for their sport, it seems that many people view the game in the same way Will Rogers did: as one that will make a cheater out of anyone. That is not terribly surprising when you stop to think about it. In golf, players hit a tiny ball across a great distance, replete with obstacles, into a very small hole. In other words, it's extremely frustrating and difficult, and when we're the ones judging our own performance, it seems that there would be many times where we might be a little extra lenient when it comes to applying the rules to our own score.

So in our quest to learn more about dishonesty, we turned to our nation's many golfers. In 2009, Scott McKenzie (a Duke undergraduate student at the time) and I carried out a study in which we asked thousands of golfers a series of questions about how they play the game and, most importantly, how they cheat. We asked them to imagine situations in which nobody could observe them (as is often the case in golf) and they could decide to follow the rules (or not) without any negative consequences. With the help of a company that manages golf courses, we e-mailed golfers around the United States, asking them to participate in a survey on golf in return for a chance to win all kinds of high-end golf equipment. About twelve thousand golfers answered our call, and here is what we learned.

Moving the Ball

"Imagine," we asked the participants, "that as the average golfer approaches their ball they realize that it would be highly advantageous if the ball would lie 4 inches away from where it is currently. How likely do you think the average golfer would be to move the ball by these 4 inches?"

This question appeared in three different versions, each describing a different approach for improving the unfortunate location of the ball (it is a curious coincidence, by the way, that in golf lingo the location of the ball is called a "lie"). How comfortable do you think the average golfer would be about moving the ball 4 inches (1) with his club; (2) with his shoe; and (3) by picking the ball up and placing it 4 inches away?

The "moving the ball" questions were designed to see whether in golf, as in our previous experiments, the distance from the dishonest action would change the tendency to behave immorally. If distance worked in the same way as the token experiment we discussed earlier (see chapter 2, "Fun with the Fudge Factor"), we would expect to have the lowest level of cheating when the movement was carried out explicitly with one's hand; we would see higher levels of cheating when the movement was accomplished with a shoe; and we would see the highest level of dishonesty when the distance was greatest and the movement was achieved via an instrument (a golf club) that removed the player from direct contact with the ball.

What our results showed is that dishonesty in golf, much as in our other experiments, is indeed directly influenced by the psychological distance from the action. Cheating becomes much simpler when there are more steps between us and the dishonest act. Our respondents felt that moving the ball with a club was the easiest, and they stated that the average golfer would do it 23 percent of the time. Next was kicking the ball (14 percent of the time), and finally, picking up and moving the ball was the most morally difficult way to improve the ball's position (10 percent of the time).

These results suggest that if we pick up the ball and reposition it, there is no way we can ignore the purposefulness and intentionality of the act, and accordingly we cannot help but feel that we have done something unethical. When we kick the ball with our shoe, there is a little bit of distance from the act, but we are still the ones doing the kicking. But when the club is doing the tapping (and especially if we move the ball in a slightly haphazard and imprecise way) we can justify what we have done relatively easily. "After all," we might say to ourselves, "perhaps there was some element of luck in exactly how the ball ended up being positioned." In that case, we can almost fully forgive ourselves.

Taking Mulligans

Legend has it that in the 1920s, a Canadian golfer named David Mulligan was golfing at a country club in Montreal. One day, he teed off and wasn't happy with his shot, so he reteed and tried again. According to the story, he called it a "correction shot," but his partners thought "mulligan" was a better name, and it stuck as the official term for a "do-over" in golf.

These days, if a shot is egregiously bad, a golfer might write it off as a "mulligan," place the ball back at its original starting point, and score himself as if the shot never happened (one of my friends refers to her husband's ex-wife as a "mulligan"). Strictly speaking, mulligans are never allowed, but in friendly games, players sometimes agree in advance that mulligans are permitted. Of course, even when mulligans are not legal nor agreed upon, golfers still take them from time to time, and those illegal mulligans were the focus of our next set of questions.

We asked our participants how likely other golfers are to take illegal mulligans when they could do it without being noticed by the other players. In one version of this question, we asked them about the likelihood of someone taking an illegal mulligan on the first hole. In the second version of the question we asked them about the likelihood of taking an illegal mulligan on the ninth hole.

To be clear, the rules don't differentiate between these two acts: they are equally prohibited. At the same time, it seems that it is easier to rationalize a do-over on the first hole than on the ninth hole. If you're on the first hole and you start over, you can pretend that "now I am really starting the game, and from now on every shot will count." But if you are on the ninth hole, there is no way for you to pretend that the game has not yet started. This means that if you take a mulligan you have to admit to yourself that you are simply not counting a shot.

As we would expect based on what we already knew about self-justification from our other experiments, we found a vast difference in the willingness to take mulligans. Our golfers predicted that 40 percent of golfers would take a mulligan on the first hole while (only?) 15 percent of golfers would take a mulligan on the ninth hole.

Fuzzy Reality

In a third set of questions, we asked the golfers to imagine that they shot 6 strokes on a par–5 hole (a hole that good players can complete in 5 strokes). In one version of this question we asked whether the average golfer would write down "5" instead of "6" on his scorecard. In the second version of this question, we asked how likely the average golfer would be to record his score accurately but then, when it comes to adding the scores up, count the 6 as a 5 and thus get the same discount on the score but doing so by adding incorrectly.

We wanted to see whether it would be more easily justifiable to write down the score wrongly to start with, because once the score is written, it is hard to justify adding incorrectly (akin to repositioning a ball by hand). After all, adding incorrectly is an explicit and deliberate act of cheating that cannot be as easily rationalized. That was indeed what we found. Our golfers predicted that in such cases, 15 percent of golfers would write down an improved score, while many fewer (5 percent) would add their score inaccurately.

The great golfer Arnold Palmer once said, "I have a tip that can take five strokes off anyone's golf game. It's called an eraser." It appears, however, that the vast majority of golfers are unwilling to go this route, or at least that they would have an easier time cheating if they did not write the score correctly from the get-go. So here's the timeless "if-a-tree-falls-in-the-forest"-type question: if a golfer shoots a 6 on a par–5 hole, the score is not recorded, and there is no one there to see it—is his score a 6 or a 5?

LYING ABOUT A score in this way has a lot in common with a classic thought experiment called "Schrödinger's cat." Erwin Schrödinger was an Austrian physicist who, in 1935, described the following scenario: A cat is sealed in a steel box with a radioactive isotope that may or may not decay. If it does decay, it will set off a chain of events that will result in the cat's death. If not, the cat will continue living. In Schrödinger's story, as long as the box remains sealed, the cat is suspended between life and death; it cannot be described as either alive or dead. Schrödinger's scenario was intended to critique an interpretation of physics that held that quantum mechanics did not describe objective reality—rather, it dealt only in probability. Leaving the philosophical aspects of physics aside for now, Schrödinger's cat story might serve us well here when thinking about golf scores. A golf score might be a lot like Schrödinger's alive-and-dead cat: until it is written down, it does not really exist in either form. Only when it's written down does it obtain the status of "objective reality."

YOU MAY BE wondering why we asked participants about "the average golfer" rather than about their own behavior on the course. The reason for this was that we expected that, like most people, our golfers would lie if they were asked directly about their own tendency to behave in unethical ways. By asking them about the behavior of others, we expected that they would feel free to tell the truth without feeling that they are admitting to any bad behavior themselves.*

Still, we also wanted to examine what unethical behaviors golfers would be willing to admit to about their own behavior. What we found was that although many "other golfers" cheat, the particular participants in our study were relative angels: when asked about their own behavior, they admitted to moving the ball with their club in order to improve their lie just 8 percent of the time. Kicking the ball with their shoe was even more rare (just 4 percent of the time), and picking up the ball and moving it occurred only 2.5 percent of the time. Now, 8 percent, 4 percent, and 2.5 percent might still look like big numbers (particularly given the fact that a golf course has 18 holes and many different ways to be dishonest), but they pale in comparison to what "other golfers" do.

We found similar differences in golfers' responses regarding mulligans and scorekeeping. Our participants re-

* Think about all the cases in which people ask for advice about how to behave in embarrassing situations—not for themselves but for a "friend."

ported that they would take a mulligan on the first hole only 18 percent of the time and on the ninth hole just 4 percent of the time. They also said that they would write in the wrong score only 4 percent of the time, and barely 1 percent copped to something as egregious as mistallying their scores.

So here's a summary of our results:

Question Type	Question	Tendency to Cheat	
		Other Golfers'	One's Own
Moving the ball	With club	23%	8%
	Kicking	14%	4%
	Picking up	10%	2.5%
Mulligans	On first hole	40%	18%
	On ninth hole	15%	4%
Recording score	Writing wrongly	15%	4%
	Adding wrongly	5%	1%

I am not sure how you want to interpret these differences, but it looks to me as though golfers not only cheat a lot in golf, they also lie about lying.

WHAT HAVE WE learned from this fairway adventure? It seems that cheating in golf captures many of the nuances we discovered about cheating in our laboratory experiments. When our actions are more distant from the execution of the dishonest act, when they are suspended, and when we can more easily rationalize them, golfers—like every other human on the planet—find it easier to be dishonest. It also seems that golfers, like everyone else, have the ability to be dishonest but at the same time think of themselves as honest. And what have we learned about the cheating of businesspeople? Well. When the rules are somewhat open to interpretation, when there are gray areas, and when people are left to score their own performance—even honorable games such as golf can be traps for dishonesty.

CHAPTER 3

Blinded by Our Own Motivations

Picture your next dental appointment. You walk in, exchange pleasantries with the receptionist, and begin leafing through some old magazines while waiting for your name to be called.

Now let's imagine that since your last visit, your dentist went out and bought an innovative and expensive piece of dental equipment. It's a dental CAD/CAM (short for computer-aided design/computer-aided manufacturing) machine, a cutting-edge device used to customize tooth restorations such as crowns and bridges. The device works in two steps. First it displays a 3D replica of the patient's teeth and gums on a computer screen, allowing the dentist to trace the exact shape of the crown—or whatever the restoration—against the screen's image. This is the CAD part. Then comes the CAM part; this device molds ceramic material into a crown according to the dentist's blueprint. Altogether, this fancy machine comes with a hefty price tag.

But let's get back to you. Just as you finish skimming an article about some politician's marital troubles and are about to start a story about the next it-girl, the receptionist calls your name. "Second room to the left," she says.

You situate yourself in the dentist's chair and engage in a bit of small talk with the hygienist, who pokes around your mouth for a while and follows up with a cleaning. Before long, your dentist walks in.

The dentist repeats the same general poking procedure, and as he checks your teeth he tells the hygienist to mark teeth 3 and 4 for further observation and to mark tooth 7 as having craze lines.

"Huh? Caze wha?" you gurgle, with your mouth open wide and the suction tube pulling on the right side of your mouth.

The dentist stops, pulls the instruments out, carefully places them on the tray next to him, and sits back in his chair. He then starts explaining your situation: "Craze lines are what we call certain small cracks in the tooth enamel. But no problem, we have a great solution for this. We'll just use the CAD/CAM to fit you with a crown, problem solved. How about it?" he asks.

You waver a little, but after you get his assurance that it won't hurt one bit, you agree. After all, you have been seeing this dentist for a long time, and although some of his treatments over the years were rather unpleasant, you feel that he has generally treated you well.

Now, I should point out—because your dentist might not—that craze lines are basically very, very small cracks in the enamel of your teeth, and what's more, they're almost always completely asymptomatic; many people have them and aren't bothered by them in the least. So, in effect, it's usually unnecessary to target craze lines with any kind of treatment.

LET ME GIVE you one real-life story from my friend Jim, the former vice president of a large dental company. Over the years, Jim has encountered his fair share of oddball dental cases, but one CAD/CAM story he told me was particularly horrible.

A few years after the CAD/CAM equipment came onto the market, one particular dentist in Missouri invested in the equipment, and from that point on he seemed to start looking at craze lines differently. "He wanted to crown everything," Jim told me. "He was excited and enthusiastic to use his brand-new gadget, so he recommended that many of his patients improve their smiles, using, of course, his state-of-the-art CAD/CAM equipment."

One of his patients was a young law student with asymptomatic craze lines; still, he recommended that she get a crown. The young woman complied, because she was used to listening to her dentist's advice, but guess what? Because of the crown, her tooth became symptomatic and then died, forcing her to go in for a root canal. But wait, it gets worse. The root canal failed and had to be redone, and that second root canal failed as well. As a result, the woman had no choice but to undergo more complex and painful surgery. So what began as a treatment for harmless craze lines ultimately resulted in a lot of pain and financial cost for this young woman.

After the woman graduated from law school, she did her homework and realized that (surprise!) she'd never needed that crown in the first place. As you can imagine, she wasn't thrilled by this, so she went after the dentist with a vengeance, took him to court, and won.

NOW, WHAT CAN we make of this tale? As we've already learned, people don't need to be corrupt in order to act in problematic and sometimes damaging ways. Perfectly well-meaning people can get tripped up by the quirks of the human mind, make egregious mistakes, and still consider themselves to be good and moral. It's safe to say that most dentists are competent, caring individuals who approach their work with the best of intentions. Yet, as it turns out, biased incentives can—and do—lead even the most upstanding professionals astray.

Think about it. When a dentist decides to purchase a new device, he no doubt believes it will help him better serve his patients. But it can also be an expensive venture. He wants to use it to improve patient care, but he also wants to recover his investment by charging his patients for using this wonderful new technology. So, consciously or not, he looks for ways to do so, and voilà! The patient ends up with a crown—sometimes necessary, other times not.

To be clear, I don't think dentists (or the vast majority of people, for that matter) carry out an explicit calculation of costs and benefits by weighing patients' well-being against their own pockets and then deliberately choose their own self-interest over their patients' best interest. Instead, I suspect that some dentists who purchase the CAD/CAM equipment are reacting to the fact that they have invested a great deal of money in the device and want to make the most of it.

This information then colors the dentists' professional judgment, leading them to make recommendations and decisions that are in their own self-interest rather than doing what is best for the patient.

You might think that instances like this, when a service provider is pulled in two directions (generally referred to as a conflict of interest), are rare. But the reality is that conflicts of interest influence our behavior in all kinds of places and, quite frequently, both professionally and personally.

Figure 2: How Conflicts of Interest Can Work on Dentists

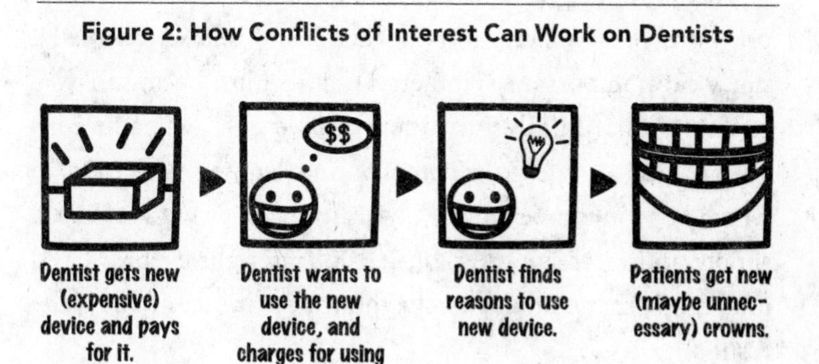

Can I Tattoo Your Face?

Some time ago I ran smack into a rather strange conflict of interest. In this case I was the patient. As a young man in my midtwenties—about six or seven years after I was originally injured*—I went back to the hospital for a routine checkup.

* When I was a teenager, a magnesium flare exploded next to me. I suffered massive third-degree burns and underwent many operations and treatments over the subsequent years. For more details, see my previous books.

On that particular visit, I met with a few physicians, and they reviewed my case. Later, I met the head of the burn department, who seemed especially happy to see me.

"Dan, I have a fantastic new treatment for you!" he exclaimed. "You see, because you have thick, dark hair, when you shave, no matter how closely you try to shave, there will always be little black dots where your hair grows. But since the right side of your face is scarred, you don't have any facial hair or small black dots on that side, making your face look asymmetrical."

At that point, he launched into a short lecture on the importance of symmetry for aesthetic and social reasons. I knew how important symmetry was to him, because I was given a similar minilecture a few years earlier, when he convinced me to undergo a complex and lengthy operation in which he would take part of my scalp together with its blood supply and re-create the right half of my right eyebrow. (I'd undergone that complex twelve-hour operation and liked the results.)

Then came his proposal: "We have started tattooing little dots resembling stubble onto scarred faces much like yours, and our patients have been incredibly happy with the results."

"That sounds interesting," I said. "Can I talk to one of the patients that had this procedure?"

"Unfortunately you can't—that would violate medical confidentiality," he said. Instead, he showed me pictures of the patients—not of their whole faces, just the parts that were tattooed. And sure enough, it did look as though the scarred faces were covered with black stubblelike specks.

But then I thought of something. "What happens when I grow old and my hair turns gray?" I asked.

"Oh, that's no problem," he replied. "When that happens, we'll just lighten up the tattoo with a laser." Satisfied, he got up, adding "Come back tomorrow at nine. Just shave the left side of your face as you usually do, with the same closeness of shave that you like to keep, and I'll tattoo the right side of your face to look the same. I guarantee that by noon, you'll be happier and more attractive."

I mulled over the possible treatment on my drive home and for the rest of the day. I also realized that in order to get the full benefit from this treatment, I would have to shave in exactly the same way for the rest of my life. I walked into the department head's office the next morning and told him that I was not interested in the procedure.

I did not expect what came next. "What is wrong with you?" he growled. "Do you like looking unattractive? Do you derive some kind of strange pleasure from looking asymmetrical? Do women feel sorry for you and give you sympathy sex? I'm offering you a chance to fix yourself in a very simple and elegant way. Why not just take it and be grateful?"

"I don't know," I said. "I'm just uncomfortable with the idea. Let me think about it some more."

You may find it hard to believe that the department head could be so aggressive and harsh, but I assure you this is exactly what he told me. At the same time, it was not his usual manner with me, so I was puzzled by his unrelenting approach. In fact, he was a fantastic, dedicated doctor who treated me well and worked very hard to make me better. It was also not the first time I refused a treatment. Over many years of interacting with medical professionals, I had decided to have some treatments and not others. But none of my doctors, including the head of the burn department, had ever tried to guilt me into having a treatment.

In an attempt to solve this mystery, I went to his deputy, a younger doctor with whom I had a friendly rapport. I asked him to explain why the department head had put me under such pressure.

"Ah, yes, yes," the deputy said. "He's already performed this procedure on two patients, and he needs just one more in order to publish a scientific paper in one of the leading medical journals."

This additional information certainly helped me better understand the conflict of interest I was up against. Here was a really good physician, someone I had known for many years and who had consistently treated me with compassion and great care. Yet, despite the fact that he cared a great deal about me in general, in this instance he was unable to see past his conflict of interest. It goes to show just how hard it is to overcome conflicts of interests once they fundamentally color our view of the world.

After years of experience publishing in academic journals myself, I now have a greater understanding of this physician's conflict of interest (more about this later). Of course, I've never tried to coerce anyone into tattooing his face—but there's still time for that.

The Hidden Cost of Favors

One other common cause of conflicts of interest is our inherent inclination to return favors. We humans are deeply social creatures, so when someone lends us a hand in some way or presents us with a gift, we tend to feel indebted. That feeling can in turn color our view, making us more inclined to try to help that person in the future.

One of the most interesting studies on the impact of favors was carried out by Ann Harvey, Ulrich Kirk, George Denfield, and Read Montague (at the time all were at the Baylor College of Medicine). In this study, Ann and her colleagues looked into whether a favor could influence aesthetic preferences.

When participants arrived at the neuroscience lab at Baylor, they were told that they would be evaluating art from two galleries, one called "Third Moon" and another called "Lone Wolfe." The participants were informed that the galleries had generously provided their payment for participating in this experiment. Some were told that their individual payment was sponsored by Third Moon, while the others were told that their individual payment was sponsored by Lone Wolfe.

Armed with this information, the participants moved to the main part of the experiment. One by one, they were asked to remain as motionless as possible in a functional magnetic resonance imagining (fMRI) scanner, a large machine with a cylinder-shaped hole in the middle. Once they were situated inside the massive magnet, they viewed a series of sixty paintings, one at a time. All the paintings were by Western artists dating from the thirteenth through the twentieth century and ranged from representational to abstract art. But the sixty paintings were not all that they saw. Near the top-left corner of each painting was the handsome logo of the gallery where that particular picture could be purchased—which meant that some pictures were presented as if they came from the gallery that sponsored the participant, and some pictures were presented as if they came from the non-sponsoring gallery.

Once the scanning portion of the experiment was over, each participant was asked to take another look at each of the painting-logo combinations, but this time they were asked to rate each of the pictures on a scale that ranged from "dislike" to "like."

With the rating information in hand, Ann and her colleagues could compare which paintings the participants liked more, the ones from Third Moon or the ones from Lone Wolfe. As you might suspect, when the researchers examined the ratings they found that participants gave more favorable ratings to the paintings that came from their sponsoring gallery.

You might think that this preference for the sponsoring gallery was due to a kind of politeness—or maybe just lip service, the way we compliment friends who invite us for dinner even when the food is mediocre. This is where the fMRI part of the study came in handy. Suggesting that the effects of reciprocity run deep, the brain scans showed the same effect; the presence of the sponsor's logo increased the activity in the parts of the participants' brains that are related to pleasure (particularly the ventromedial prefrontal cortex, a part of the brain that is responsible for higher-order thinking, including associations and meaning). This suggested that the favor from the sponsoring gallery had a deep effect on how people responded to the art. And get this: when participants were asked if they thought that the sponsor's logo had any effect on their art preferences, the universal answer was "No way, absolutely not."

What's more, different participants were given varying

amounts of money for their time in the experiments. Some received $30 from their sponsoring gallery, others received $100. At the highest level, participants were paid $300. It turned out that the favoritism toward the sponsoring gallery increased as the amount of earnings grew. The magnitude of brain activation in the pleasure centers of the brain was lowest when the payment was $30, higher when the payment was $100, and highest when the payment was $300.

These results suggest that once someone (or some organization) does us a favor, we become partial to anything related to the giving party—and that the magnitude of this bias increases as the magnitude of the initial favor (in this case the amount of payment) increases. It's particularly interesting that financial favors could have an influence on one's preferences for art, especially considering that the favor (paying for their participation in the study) had nothing at all to do with the art, which had been created independently of the galleries. It is also interesting to note that participants knew the gallery would pay their compensation regardless of their ratings of the paintings and yet the payment (and its magnitude) established a sense of reciprocity that guided their preferences.

Fun with Pharma

Some people and companies understand this human propensity for reciprocity very well and consequently spend a lot of time and money trying to engender a feeling of obligation in others. To my mind, the profession that most embodies this type of operation, that is, the one that depends most on creating conflicts of interests, is—of course—that of governmental lobbyists, who spend a small fraction of their time informing politicians about facts as reported by their employers and the rest of their time trying to implant a feeling of obligation and reciprocity in politicians who they hope will repay them by voting with their interest in mind.

But lobbyists are not alone in their relentless pursuit of conflicts of interest, and some other professions could arguably give them a run for their well-apportioned money. For example, let's consider the way representatives for drug companies (pharma reps) run their business. A pharma rep's job is to visit doctors and convince them to purchase medical equipment and drugs to treat everything from A(sthma)to Z(ollinger-Ellison syndrome). First they may give a doctor a free pen with their logo, or perhaps a notepad, a mug, or maybe some free drug samples. Those small gifts can subtly influence physicians to prescribe a drug more often—all because they feel the need to give back.[1]

But small gifts and free drug samples are just a few of the many psychological tricks that pharma reps use as they set out to woo physicians. "They think of everything," my friend and colleague (let's call him MD) told me. He went on to explain that drug companies, especially smaller ones, train their reps to treat doctors as if they were gods. And they seem to have a disproportionately large reserve of attractive reps. The whole effort is coordinated with military precision. Every self-respecting rep has access to a database that tells them exactly what each doctor has prescribed over the last quarter (both that company's drugs as well as their competitors'). The reps also make it their business to know what kind of food each doctor and their office staff likes, what time of day they are most likely to see reps, and also which type of rep gets the most face time with the doctors. If the doctor is noted to spend more time with a certain female rep, they may adjust that rep's rotation so that she can spend more time in that office. If the doctor is a fan of the military, they'll send him a veteran. The reps also make it a point to be agreeable with the doctor's outer circles, so when the rep arrives they start by handing out candy and other small gifts to the nurses and the front desk, securing themselves in everyone's good graces from the get-go.

One particularly interesting practice is the "dine-and-dash," where, in the name of education, doctors can simply pull up at prespecified take-out restaurants and pick up whatever they want. Even medical students and trainees are pulled into some schemes. One particularly creative example of this strategy was the famous black mug. A black mug with the company's logo was handed out to doctors and residents, and the company arranged it such that a doctor could take this mug to any location of a local coffee chain (which shall go unnamed) and get as much espresso or cappuccino as he or she wanted. The clamor for this mug was so great that it became a status symbol among students and trainees. As these practices became more extravagant, there was also more regulation from hospitals and the American Medical Association, limiting the use of these aggressive marketing tactics. Of course, as the regulations become more stringent, pharma reps continue to search for new and innovative approaches to influence physicians. And the arms race continues . . .*

---

* Perhaps the most telling evidence for the pharma industry's influence is the fact that my insider for this interview insisted that I keep his name confidential to avoid being blacklisted by pharma.

A FEW YEARS AGO, my colleague Janet Schwartz (a professor at Tulane University) and I invited some pharmaceutical reps to dinner. We basically tried the pharma reps at their own game; we took them to a nice restaurant and kept the wine flowing. Once we had them feeling happily lubricated, they were ready to tell us the tricks of their trade. And what we learned was fairly shocking.

Picture one of those pharma reps, an attractive, charming man in his early twenties. Not the kind of guy who would have any trouble finding a date. He told us how he had once persuaded a reluctant female physician to attend an informational seminar about a medication he was promoting—by agreeing to escort her to a ballroom dancing class. It was an unstated quid pro quo: the rep did a personal favor for the doctor, and the doctor took his free drug samples and promoted the product to her patients.

Another common practice, the reps told us, was to take fancy meals to the entire doctor's office (one of the perks of being a nurse or receptionist, I suppose). One doctor's office even required alternating days of steak and lobster for lunch if the reps wanted access to the doctors. Even more shocking, we found out that physicians sometimes called the reps into the examination room (as an "expert") to directly inform patients about the way certain drugs work.

Hearing stories from the reps who sold medical devices was even more disturbing. We learned that it's common practice for device reps to peddle their medical devices in the operating room in real time and while a surgery is under way.

Janet and I were surprised at how well the pharmaceutical reps understood classic psychological persuasion strategies and how they employed them in a sophisticated and intuitive manner. Another clever tactic that they told us about involved hiring physicians to give a brief lecture to other doctors about a drug they were trying to promote. Now, the pharma reps really didn't care about what the audience took from the lecture—what they were actually interested in was the effect that giving the lecture had on the speaker. They found that after giving a short lecture about the benefits of a certain drug, the speaker would begin to believe his own words and soon prescribe accordingly. Psychological studies show that we quickly and easily start believing whatever comes out of our own mouths, even when the original reason for expressing the opinion is no longer relevant (in the doctors' case, that they were paid to say it). This is cognitive dissonance at play; doctors reason that if they are telling others about a drug, it must be good—and so their own beliefs change to correspond to their speech, and they start prescribing accordingly.

The reps told us that they employed other tricks too, turning into chameleons—switching various accents, personalities, and political affiliations on and off. They prided themselves on their ability to put doctors at ease. Sometimes a collegial relationship expanded into the territory of social friendship—some reps would go deep-sea fishing or play basketball with the doctors as friends. Such shared experiences allowed the physicians to more happily write prescriptions that benefited their "buddies." The physicians, of course, did not see that they were compromising their values when they were out fishing or shooting hoops with the drug reps; they were just taking a well-deserved break with a friend with whom they just happened to do business. Of course, in many cases the doctors probably didn't realize that they were being manipulated—but there is no doubt that they were.

DISGUISED FAVORS ARE one thing, but there are many cases when conflicts of interest are more easily recognizable. Sometimes a drug maker pays a doctor thousands of dollars in consulting fees. Sometimes the company donates a building or gives an endowment to a medical researcher's department in the hope of influencing his views. This type of action creates immense conflicts of interest—especially at medical schools, where pharmaceutical bias can be passed from the medical professor to medical students and along to patients.

Duff Wilson, a reporter for *The New York Times*, described one example of this type of behavior. A few years ago, a Harvard Medical School student noticed that his pharmacology professor was heavily promoting the benefits of cholesterol drugs and downplaying their side effects. When the student did some Googling, he discovered that the professor was on the payroll of ten drug companies, five of which made cholesterol drugs. And the professor wasn't alone. As Wilson put it, "Under the school's disclosure rules, about 1,600 of 8,900 professors and lecturers at Harvard Medical School have reported to the dean that they or a family member had a financial interest in a business related to their teaching, research, or clinical care."[2] When professors publicly pass drug recommendations off as academic knowledge, we have a serious problem.

Fudging the Numbers

If you think that the world of medicine is rife with conflicts of interest, let's consider another profession in which these conflicts may be even more widespread. Yes, I'm talking about the wonderland of financial services.

Say it's 2007, and you've just accepted a fantastic banking job on Wall Street. Your bonus could be in the neighborhood of $5 million a year, but only if you view mortgage-backed securities (or some other new financial instrument) in a positive light. You're being paid a lot of money to maintain a distorted view of reality, but you don't notice the tricks that your big bonus plays on your perception of reality. Instead, you are quickly convinced that mortgage-backed securities are every bit as solid as you want to believe they are.

Once you've accepted that mortgage-backed securities are the wave of the future, you're at least partially blind to their risks. On top of that, it's notoriously hard to evaluate how much securities are really worth. As you sit there with your large and complex Excel spreadsheet full of parameters and equations, you try to figure out the real value of the securities. You change one of the discount parameters from 0.934 to 0.936, and right off the bat you see how the value of the securities jumps up. You continue to play around with the numbers, searching for parameters that provide the best representation of "reality," but with one eye you also see the consequences of your parameter choices for your personal financial future. You continue to play with the numbers for a while longer, until you are convinced that the numbers truly represent the ideal way to evaluate mortgage-backed securities. You don't feel bad because you are certain that you have done your best to represent the values of the securities as objectively as possible.

Moreover, you aren't dealing with real cash; you are only playing with numbers that are many steps removed from cash. Their abstractness allows you to view your actions more as a game, and not as something that actually affects people's homes, livelihoods, and retirement accounts. You are also not alone. You realize that the smart financial engineers in the offices next to yours are behaving more or less the same way as you and when you compare your evaluations to theirs, you realize that a few of your coworkers have chosen even more extreme values than yours. Believing that you are a rational creature, and believing that the market is always correct, you are even more inclined to accept what you're doing—and what everyone else is doing (we'll learn more about this in chapter 8)—as the right way to go. Right?

Of course, none of this is actually okay (remember the financial crisis of 2008?), but given the amount of money involved, it feels natural to fudge things a bit. And it's perfectly human to behave this way. Your actions are highly problematic, but you don't see them as such. After all, your conflicts of interest are supported by the facts that you're not dealing with real money; that the financial instruments are mind-bogglingly complex; and that every one of your colleagues is doing the same thing.

The riveting (and awfully distressing) Academy Award–winning documentary *Inside Job* shows in detail how the financial services industry corrupted the U.S. government, leading to a lack of oversight on Wall Street and to the financial meltdown of 2008. The film also describes how the financial services industry paid leading academics (deans, heads of departments, university professors) to write expert reports in the service of the financial industry and Wall

Street. If you watch the film, you will most likely feel puzzled by the ease with which academic experts seemed to sell out, and think that you would never do the same.

But before you put a guarantee on your own standards of morality, imagine that I (or you) were paid a great deal to be on Giantbank's audit committee. With a large part of my income depending on Giantbank's success, I would probably not be as critical as I am currently about the bank's actions. With a hefty enough incentive I might not, for example, repeatedly say that investments must be transparent and clear and that companies need to work hard to try to overcome their conflicts of interests. Of course, I've yet to be on such a committee, so for now it's easy for me to think that many of the actions of the banks have been reprehensible.

Academics Are Conflicted Too

When I reflect on the ubiquity of conflicts of interest and how impossible they are to recognize in our own lives, I have to acknowledge that I'm susceptible to them as well.

We academics are sometimes called upon to use our knowledge as consultants and expert witnesses. Shortly after I got my first academic job, I was invited by a large law firm to be an expert witness. I knew that some of my more established colleagues provided expert testimonials as a regular side job for which they were paid handsomely (though they all insisted that they didn't do it for the money). Out of curiosity, I asked to see the transcripts of some of their old cases, and when they showed me a few I was surprised to discover how one-sided their use of the research findings was. I was also somewhat shocked to see how derogatory they were in their reports about the opinions and qualifications of the expert witnesses representing the other side—who in most cases were also respectable academics.

Even so, I decided to try it out (not for the money, of course), and I was paid quite a bit to give my expert opinion.* Very early in the case I realized that the lawyers I was working with were trying to plant ideas in my mind that would buttress their case. They did not do it forcefully or by saying that certain things would be good for their clients. Instead, they asked me to describe all the research that was relevant to the case. They suggested that some of the less favorable findings for their position might have some methodological flaws and that the research supporting their view was very important and well done. They also paid me warm compliments each time that I interpreted research in a way that was useful to them. After a few weeks, I discovered that I rather quickly adopted the viewpoint of those who were paying me. The whole experience made me doubt whether it's at all possible to be objective when one is paid for his or her opinion. (And now that I am writing about my lack of objectivity, I am sure that no one will ever ask me to be an expert witness again—and maybe that's a good thing.)

The Drunk Man and the Data Point

I had one other experience that made me realize the dangers of conflicts of interest; this time it was in my own research.

---

* This was the first time that I was paid a lot by the hour, and I was intrigued by how I started to view many decisions in terms of "work hours." I figured that for one hour of work I could buy a really fancy dinner and that for a few more I could buy a new bicycle. I suspect that this is an interesting way to think about what we should and should not purchase, and one day I might look into this.

At the time, my friends at Harvard were kind enough to let me use their behavioral lab to conduct experiments. I was particularly interested in using their facility because they recruited residents from the surrounding area rather than relying only on students.

One particular week, I was testing an experiment on decision making, and, as is usually the case, I predicted that the performance level in one of the conditions would be much higher than the performance level in the other condition. That was basically what the results showed—aside from one person. This person was in the condition I expected to perform best, but his performance was much worse than everyone else's. It was very annoying. As I examined his data more closely, I discovered that he was about twenty years older than everyone else in the study. I also remembered that there was one older fellow who was incredibly drunk when he came to the lab.

The moment I discovered that the offending participant was drunk, I realized that I should have excluded his data in the first place, given that his decision-making ability was clearly compromised. So I threw out his data, and instantly the results looked beautiful—showing exactly what I expected them to show. But, a few days later I began thinking about the process by which I decided to eliminate the drunk guy. I asked myself: what would have happened if this fellow had been in the other condition—the one I expected to do worse? If that had been the case, I probably would not have noticed his individual responses to start with. And if I had, I probably would not have even considered excluding his data.

In the aftermath of the experiment, I could easily have told myself a story that would excuse me from using the drunk guy's data. But what if he hadn't been drunk? What if he had some other kind of impairment that had nothing to do with drinking? Would I have invented another excuse or logical argument to justify excluding his data? As we will see in chapter 7, "Creativity and Dishonesty," creativity can help us justify following our selfish motives while still thinking of ourselves as honest people.

I decided to do two things. First, I reran the experiment to double-check the results, which worked out beautifully. Then I decided it was okay to create standards for excluding participants from an experiment (that is, we wouldn't test drunks or people who couldn't understand the instructions). But the rules for exclusion have to be made up front, before the experiment takes place, and definitely not after looking at the data.

What did I learn? When I was deciding to exclude the drunk man's data, I honestly believed I was doing so in the name of science—as if I were heroically fighting to clear the data so that the truth could emerge. It didn't occur to me that I might be doing it for my own self-interest, but I clearly had another motivation: to find the results I was expecting. More generally, I learned—again—about the importance of establishing rules that can safeguard ourselves from ourselves.

Disclosure: A Panacea?

So what is the best way to deal with conflicts of interest? For most people, "full disclosure" springs to mind. Following the same logic as "sunshine policies," the basic assumption underlying disclosure is that as long as people publicly declare exactly what they are doing, all will be well. If professionals were to simply make their incentives clear and known to their clients, so the thinking goes, the clients can then decide for themselves how much to rely on their (biased) advice and then make more informed decisions.

If full disclosure were the rule of the land, doctors would inform their patients when they own the equipment required for the treatments they recommend. Or when they are paid to consult for the manufacturer of the drugs that they are about to prescribe. Financial advisers would inform their clients about all the different fees, payments, and commissions they get from various vendors and investment houses. With that information in hand, consumers should be able to appropriately discount the opinions of those professionals and make better decisions. In theory, disclosure seems to be a fantastic solution; it both exonerates the professionals who are acknowledging their conflicts of interest and it provides their clients with a better sense of where their information is coming from.

HOWEVER, IT TURNS out that disclosure is not always an effective cure for conflicts of interest. In fact, disclosure can sometimes make things worse. To explain how, allow me to run you through a study conducted by Daylian Cain (a professor at Yale University), George Loewenstein (a professor at Carnegie Mellon University), and Don Moore (a professor at the University of California, Berkeley). In this experiment, participants played a game in one of two roles. (By the way, what researchers call a "game" is not what any reasonable kid would consider a game.) Some of the participants played the role of estimators: their task was to guess the total amount of money in a large jar full of loose change as accurately as possible. These players were paid according to how close their guess was to the real value of the money in the jar. The closer their estimates were, the more money they received, and it didn't matter if they missed by overestimating or underestimating the true value.

The other participants played the role of advisers, and their task was to advise the estimators on their guesses. (Think of someone akin to your stock adviser, but with a much simpler task.) There were two interesting differences between the estimators and the advisers. The first was that whereas the estimators were shown the jar from a distance for a few seconds, the advisers had more time to examine it, and they were also told that the amount of money in the jar was between $10 and $30. That gave the advisers an informational edge. It made them relative experts in the field of estimating the jar's value, and it gave the estimators a very good reason to rely on their advisers' reports when formulating their guesses (comparable to the way we rely on experts in many areas of life).

The second difference concerned the rule for paying the advisers. In the control condition, the advisers were paid according to the accuracy of the estimators' guesses, so no conflicts of interest were involved. In the conflict-of-interest condition, the advisers were paid more as the estimators overguessed the value of the coins in the jar to a larger degree. So if the estimators overguessed by $1, it was good for the advisers—but it was even better if they overguessed by $3 or $4. The higher the overestimation, the less the estimator made but the more the adviser pocketed.

So what happened in the control condition and in the

conflict-of-interest condition? You guessed it: in the control condition, advisers suggested an average value of $16.50, while in the conflict-of-interest condition, the advisers suggested an estimate that was over $20. They basically goosed the estimated value by almost $4. Now, you can look at the positive side of this result and tell yourself, "Well, at least the advice was not $36 or some other very high number." But if that is what went through your mind, you should consider two things: first, that the adviser could not give clearly exaggerated advice because, after all, the estimator did see the jar. If the value had been dramatically too high, the estimator would have dismissed the suggestion altogether. Second, remember that most people cheat just enough to still feel good about themselves. In that sense, the fudge factor was an extra $4 (or about 25 percent of the amount).

The importance of this experiment, however, showed up in the third condition—the conflict-of-interest-plus-disclosure condition. Here the payment for the adviser was the same as it was in the conflict-of-interest condition. But this time the adviser had to tell the estimator that he or she (the adviser) would receive more money when the estimator overguessed. The sunshine policy in action! That way, the estimator could presumably take the adviser's biased incentives into account and discount the advice of the adviser appropriately. Such a discount of the advice would certainly help the estimator, but what about the effect of the disclosure on the advisers? Would the need to disclose eliminate their biased advice? Would disclosing their bias stretch the fudge factor? Would they now feel more comfortable exaggerating their advice to an even greater degree? And the billion-dollar question is this: which of these two effects would prove to be larger? Would the discount that the estimator applied to the adviser's advice be smaller or larger than the extra exaggeration of the adviser?

The results? In the conflict-of-interest-plus-disclosure condition, the advisers increased their estimates by another $4 (from $20.16 to $24.16). And what did the estimators do? As you can probably guess, they did discount the estimates, but only by $2. In other words, although the estimators did take the advisers' disclosure into consideration when formulating their estimates, they didn't subtract nearly enough. Like the rest of us, the estimators didn't sufficiently recognize the extent and power of their advisers' conflicts of interest.

The main takeaway is this: disclosure created even greater bias in advice. With disclosure the estimators made less money and the advisers made more. Now, I am not sure that disclosure will always make things worse for clients, but it is clear that disclosure and sunshine policies will not always make things better.

So What Should We Do?

Now that we understand conflicts of interest a bit better, it should be clear what serious problems they cause. Not only are they ubiquitous, but we don't seem to fully appreciate their degree of influence on ourselves and on others. So where do we go from here?

One straightforward recommendation is to try to eradicate conflicts of interest altogether, which of course is easier said than done. In the medical domain, that would mean, for example, that we would not allow doctors to treat or test their own patients using equipment that they own. Instead, we'd have to require that an independent entity, with no ties to the doctors or equipment companies, conduct the treatments and tests. We would also prohibit doctors from consulting for drug companies or investing in pharmaceutical stocks. After all, if we don't want doctors to have conflicts of interest, we need to make sure that their income doesn't depend on the number and types of procedures or prescriptions they recommend. Similarly, if we want to eliminate conflicts of interest for financial advisers, we should not allow them to have incentives that are not aligned with their clients' best interests—no fees for services, no kickbacks, and no differential pay for success and failure.

Though it is clearly important to try to reduce conflicts of interest, it is not easy to do so. Take contractors, lawyers, and car mechanics, for example. The way these professionals are paid puts them into terrible conflicts of interest because they both make the recommendation and benefit from the service, while the client has no expertise or leverage. But stop for a few minutes and try to think about a compensation model that would not involve any conflicts of interest. If you are taking the time to try to come up with such an approach, you most likely agree that it is very hard—if not impossible—to pull off. It is also important to realize that although conflicts of interest cause problems, they sometimes happen for good reason. Take the case of physicians (and dentists) ordering treatments that use equipment they own. Although this is a potentially dangerous practice from the perspective of conflicts of interest, it also has some built-in advantages: professionals are more likely to purchase equipment that they believe in; they are likely to become experts in using it; it can be much more convenient for the patient; and the doctors might even conduct some research that could help improve the equipment or the ways in which it is used.

The bottom line is that it is no easy task to come up with compensation systems that don't inherently involve—and sometimes rely on—conflicts of interest. Even if we could eliminate all conflicts of interest, the cost of doing so in terms of decreased flexibility and increased bureaucracy and oversight might not be worth it—which is why we should not overzealously advocate draconian rules and restrictions (say, that physicians can never talk to pharma reps or own medical equipment). At the same time, I do think it's important for us to realize the extent to which we can all be blinded by our financial motivations. We need to acknowledge that situations involving conflicts of interest have substantial disadvantages and attempt to thoughtfully reduce them when their costs are likely to outweigh their benefits.

As you might expect, there are many straightforward instances where conflicts of interest should simply be eliminated. For example, the conflicts for financial advisers who receive side payments, auditors who serve as consultants to the same firms, financial professionals who are paid handsome bonuses when their clients make money but lose nothing when their clients lose their shirts, rating agencies that are paid by the companies they rate, and politicians who accept money and favors from corporations and lobbyists in exchange for their votes; in all of these cases it seems to me that we must do our best to eradicate as many conflicts of interest as possible—most likely by regulation.

You're probably skeptical that regulation of this sort could ever happen. When regulation by the government or by professional organizations does not materialize, we as consumers should recognize the danger that conflicts of interest bring with them and do our best to seek service providers who have fewer conflicts of interest (or, if possible, none). Through the power of our wallets we can push service providers to meet a demand for reduced conflicts of interest.

Finally, when we face serious decisions in which we realize that the person giving us advice may be biased—such as when a physician offers to tattoo our faces—we should spend just a little extra time and energy to seek a second opinion from a party that has no financial stake in the decision at hand.

CHAPTER 4

Why We Blow It When We're Tired

Imagine yourself at the end of a really long, hard day. Let's say it's the most exhausting of days: moving day. You're completely exhausted. Even your hair feels tired. Cooking is certainly out of the question. You don't even have the energy to locate a pan, plate, and fork, much less put them to use. Clearly it's going to be a take-out night.

Within a block of your new place are three restaurants. One is a little bistro with fresh salads and paninis. Another is a Chinese place; the greasy, salty smells emanating from within make the back of your mouth tingle. There's also a cute mom-and-pop pizzeria where the locals enjoy cheesy slices twice the size of their faces. To which restaurant do you drag your tired, aching body? Which kind of cuisine would you prefer to enjoy on your new floor? By contrast, consider what your choice might be if the meal were after an afternoon spent relaxing in the backyard with a good book.

In case you haven't noticed, on stressful days many of us give in to temptation and choose one of the less healthy alternatives. Chinese takeout and pizza are practically synonymous with moving day, conjuring up an image of a young, attractive, tired, but happy couple surrounded by cardboard boxes and eating chow mein out of the box with chopsticks. And we all remember the times college friends offered us pizza and beer in exchange for helping them move.

This mysterious connection between exhaustion and the consumption of junk food is not just a figment of your imagination. And it is the reason why so many diets die on the chopping block of stress and why people start smoking again after a crisis.

Let Us Eat Cake

The key to this mystery has to do with the struggle between the impulsive (or emotional) and the rational (or deliberative) parts of ourselves. This is not a new idea; many seminal books (and academic papers) throughout history have had something to say about the conflicts between desire and reason. We have Adam and Eve, tempted by the prospect of forbidden knowledge and that succulent fruit. There was Odysseus, who knew he'd be lured by the Sirens' song and cleverly ordered his crew to tie him to the mast and fill their ears with wax to muffle the tantalizing call (that way, Odysseus could have it both ways—he could hear the song without worrying that the men would wreck the ship). And in one of the most tragic struggles between emotion and reason, Shakespeare's Romeo and Juliet fell hard for each other, despite Friar Laurence's warning that untamed passion only brings disaster.

In a fascinating demonstration of the tension between reason and desire, Baba Shiv (a professor at Stanford University) and Sasha Fedorikhin (a professor at Indiana University) examined the idea that people fall into temptation more frequently when the part of their brain that is in charge of deliberative thinking is otherwise occupied. To reduce participants' ability to think effectively, Baba and Sasha did not remove parts of their brains (as animal researchers sometimes do), nor did they use magnetic pulses to disrupt thinking (though there are machines that can do that). Instead, they decided to tax their participants' ability to think by piling on what psychologists call cognitive load. Simply put, they wanted to find out whether having a lot on one's mind would leave less cognitive room for resisting temptation and make people more likely to succumb to it.

Baba and Sasha's experiment went like this: they divided participants into two groups and asked members of one group to remember a two-digit number (something like, say, 35) and they asked members of the other group to remember a seven-digit number (say, 7581280). The participants were told that in order to get their payment for the experiment, they would have to repeat the number to another experimenter who was waiting for them in a second room at the other end of the corridor. And if they didn't remember the number? No reward.

The participants lined up to take part in the experiment and were briefly shown either the two-digit number or the seven-digit number. With their numbers in mind, they each walked down the hall to the second room where they would be asked to recall the number. But on the way, they unexpectedly passed by a cart displaying pieces of rich, dark chocolate cake and bowls of colorful, healthy-looking fruit. As participants passed the cart, another experimenter told them that once they got to the second room and recited their number they could have one of the two snacks—but they had to make their decision right then, at the cart. The participants made their choice, received a slip of paper indicating their chosen snack, and off they went to the second room.

What decisions did participants make while laboring under more and less cognitive strain? Did the "Yum, cake!" impulse win the day, or did they select the healthy fruit salad (the well-reasoned choice)? As Baba and Sasha suspected, the answer depended in part on whether the participants were thinking about an easy-to-remember number or a hard one. Those breezing down the hall with a mere "35" on their minds chose the fruit much more frequently than those struggling with "7581280." With their higher-level faculties preoccupied, the seven-digit group was less able to overturn their instinctive desires, and many of them ended up succumbing to the instantly gratifying chocolate cake.

The Tired Brain

Baba and Sasha's experiment showed that when our deliberative reasoning ability is occupied, the impulsive system gains more control over our behavior. But the interplay between our ability to reason and our desires gets even more complicated when we think about what Roy Baumeister (a professor at Florida State University) coined "ego depletion."

To understand ego depletion, imagine that you're trying to lose a few extra pounds. One day at work, you are eyeing a cheese danish at the morning meeting, but you're trying to be good, so you work very hard to resist the temptation and just sip your coffee instead. Later that day, you are craving fettuccine alfredo for lunch but you force yourself to order a garden salad with grilled chicken. An hour later, you want to knock off a little early since your boss is out, but you stop yourself and say, "No, I must finish this project." In each of these instances your hedonic instincts prompt you toward pleasurable types of gratification, while your laudable self-control (or willpower) applies opposing force in an attempt to counteract these urges.

The basic idea behind ego depletion is that resisting temptation takes considerable effort and energy. Think of your willpower as a muscle. When we see fried chicken or a chocolate milkshake, our first reaction is an instinctive "Yum, want!" Then, as we try to overcome the desire, we expend a bit of energy. Each of the decisions we make to avoid temptation takes some degree of effort (like lifting a weight once), and we exhaust our willpower by using it over and over (like lifting a weight over and over). This means that after a long day of saying "no" to various and sundry temptations, our capacity for resisting them diminishes—until at some point we surrender and end up with a belly full of cheese danish, Oreos, french fries, or whatever it is that makes us salivate. This, of course, is a worrisome thought. After all, our days are increasingly full of decisions, along with a never-ending barrage of temptations. If our repeated attempts to control ourselves deplete our ability to do so, is it any wonder that we so often fail? Ego depletion also helps explain why our evenings are particularly filled with failed attempts at self-control—after a long day of working hard to be good, we get tired of it all. And as night falls, we are particularly likely to succumb to our desires (think of late-night snacking as the culmination of a day's worth of resisting temptation).

WHEN JUDGES GET TIRED

In case you've got a parole hearing coming up, make sure it's first thing in the morning or right after lunchtime. Why? According to a study by Shai Danziger (a professor at Tel Aviv University), Jonathan Levav (a professor at Stanford University), and Liora Avnaim-Pesso (a professor at Ben-Gurion University of the Negev), judges on parole boards tend to grant parole more frequently when they are most refreshed. Investigating a large set of parole rulings in Israel, the researchers found that parole boards were more likely to grant parole during their first cases of the day and just after their lunch breaks. Why? The default decision of parole boards is not to grant parole. But it seems that when the judges felt rejuvenated, which was first thing in the morning or after just having eaten and taken a break, they had an increased ability to override their standard decision, make a more effortful decision, and grant parole more frequently. But over the many difficult decisions of the day, as their cognitive burden

was building up, they opted for the simpler, default decision of not granting parole.

I think that PhD students (a slightly different sort of prisoner) instinctively understand this mechanism, which is why they often bring doughnuts, muffins, and cookies to their dissertation proposals and defenses. Based on the results of the parole study, it is likely that their judges are more likely to grant them academic parole and let them start their own independent lives.

Testing the Moral Muscle

In the TV series *Sex and the City*, Samantha Jones (the blondest and most salacious one, for those not in the know) finds herself in a committed relationship. She begins eating compulsively and consequently gains weight. What's interesting is the reason behind this baffling behavior. Samantha notices that her eating compulsion started when a good-looking man moved in next door—just the kind of man she would have gone after when she was single. She realizes that she's using food as a bulwark against temptation: "I eat so I don't cheat," she explains to her friends. Fictional Samantha is depleted, just like a real person. She can't resist all temptation, so she compromises by falling for food instead of promiscuity.

Sex and the City is no cinematic or psychological masterpiece, but it poses an interesting question: Might people who overtax themselves in one domain end up being less moral in others? Does depletion lead us to cheat? That is what Nicole Mead (a professor at Católica-Lisbon), Roy Baumeister, Francesca Gino, Maurice Schweitzer (a professor at the University of Pennsylvania), and I decided to check out. What would happen to real-life Samanthas who were depleted by one task and then given an opportunity to cheat on another? Would they cheat more? Less? Would they predict that they are more likely to succumb to temptation and therefore try to avoid the tempting situation altogether?

Our first experiment included several steps. First, we split our participants into two groups. We asked one group to write a short essay about what they had done the previous day without using the letters "x" and "z." To get a feeling for this task, try it yourself: In the space below, write a short synopsis of one of your favorite books, but don't use the letters "x" and "z." Note: you cannot simply omit the letters from the words—you must use words that do not contain an "x" or "z" (e.g., "bicycle").

____

____

____

____

____

____

We called this the nondepleting condition because, as you can tell, it's pretty easy to write an essay without using the letters "x" and "z."

We asked the other group to do the same thing but told them not to use the letters "a" and "n." To get a better grasp of how this version of the task is different, try writing a short synopsis of one of your favorite movies while not using any words that contain the letters "a" and "n."

As you probably discovered from your experience with the second task, trying to tell a story without using "a" and "n" required our storytellers to constantly repress the words that naturally popped into their minds. You can't write that the characters "went for a walk in the park" or "ran into each other at a restaurant."

All of those little acts of repression add up to greater depletion.

Once our participants turned in their essays, we asked them to perform a separate task for a different study, which was the main focus of this experiment. The other task was our standard matrix test.

How did things turn out? In the two control conditions, we found that both the depleted and nondepleted folks showed an equal ability to solve the math problems—which means that depletion did not diminish their basic ability to do the math. But in the two shredder conditions (in which they could cheat), things went differently. Those who wrote essays without the letters "x" and "z" and later shredded their answers indulged in a little bit of cheating, claiming to solve about one extra matrix correctly. But the participants in the shredder condition who'd undergone the ordeal of writing essays without the letters "a" and "n" took the proverbial cake: they claimed to have correctly solved about three extra matrices. As it turned out, the more taxing and depleting the task, the more participants cheated.

What do these findings suggest? Generally speaking, if you wear down your willpower, you will have considerably more trouble regulating your desires, and that difficulty can wear down your honesty as well.

Dead Grannies

Over the course of many years of teaching, I've noticed that there typically seems to be a rash of deaths among students' relatives at the end of the semester, and it happens mostly in the week before final exams and before papers are due. In an average semester, about 10 percent of my students come to me asking for an extension because someone has died—usually a grandmother. Of course I find it very sad and am always ready to sympathize with my students and give them more time to complete their assignments. But the question remains: what is it about the weeks before finals that is so dangerous to students' relatives?

Most professors encounter the same puzzling phenomenon, and I'll guess that we have come to suspect some kind of causal relationship between exams and sudden deaths among grandmothers. In fact, one intrepid researcher has successfully proven it. After collecting data over several years, Mike Adams (a professor of biology at Eastern Connecticut State University) has shown that grandmothers are ten times more likely to die before a midterm and nineteen times more likely to die before a final exam. Moreover, grandmothers of students who aren't doing so well in class are at even higher risk—students who are failing are fifty times more likely to lose a grandmother compared with non-failing students.

In a paper exploring this sad connection, Adams speculates that the phenomenon is due to intrafamilial dynamics, which is to say, students' grandmothers care so much about their grandchildren that they worry themselves to death over the outcome of exams. This would indeed explain why fatalities occur more frequently as the stakes rise, especially in cases where a student's academic future is in peril. With this finding in mind, it is rather clear that from a public policy perspective, grandmothers—particularly those of failing students—should be closely monitored for signs of ill health during the weeks before and during finals. Another recommendation is that their grandchildren, again particularly the ones who are not doing well in class, should not tell their grandmothers anything about the timing of the exams or how they are performing in class.

Though it is likely that intrafamilial dynamics cause this tragic turn of events, there is another possible explanation for the plague that seems to strike grandmothers twice a year. It may have something to do with students' lack of preparation and their subsequent scramble to buy more time than with any real threat to the safety of those dear old women. If that is the case, we might want to ask why it is that students become so susceptible to "losing" their grandmothers (in e-mails to professors) at semesters' end.

Perhaps at the end of the semester, the students become so depleted by the months of studying and burning the candle at both ends that they lose some of their morality and in the process also show disregard for their grandmothers' lives. If the concentration it takes to remember a longer digit can send people running for chocolate cake, it's not hard to imagine how dealing with months of cumulative material from several classes might lead students to fake a dead grandmother in order to ease the pressure (not that that's an excuse for lying to one's professors).

Just the same, to all grandmothers out there: take care of yourselves at finals time.

Red, Green, and Blue

We've learned that depletion takes away some of our reasoning powers and with them our ability to act morally.

Still, in real life we can choose to remove ourselves from situations that might tempt us to behave immorally. If we are even somewhat aware of our propensity to act dishonestly when depleted, we can take this into account and avoid temptation altogether. (For example, in the domain of dieting, avoiding temptation could mean that we decide not to shop for groceries when we're starving.)

In our next experiment, our participants could choose whether or not to put themselves into a position that would tempt them to cheat in the first place. Once again, we wanted to create two groups: one depleted, the other not. This time, however, we used a different method of mental exhaustion called the Stroop task.

In this task, we presented participants with a table of color names containing five columns and fifteen rows (for a total of seventy-five words). The words in the table were color names—red, green, and blue—printed in one of these three colors and organized in no particular order. Once the list was in front of the participants, we asked them to say the color of each word on the list aloud. Their instructions were simple: "If a word is written in red ink, regardless of what the word is, you should say 'red.' If a word is written in green ink, regardless of what the word is, you should say 'green.' And so on. Do this as fast as you can. If at any point you make a mistake, please repeat the word until you get it right."

For the participants in the nondepleting condition, the list of colors was structured such that the name of each color (e.g., green) was written in the same color of ink (green). The participants in the depleting condition were given the same instructions, but the list of words had one key difference—the ink color did not match the name of the color (for instance, the word "blue" would be printed in green ink, and the participants were asked to say "green").

To try the nondepleting condition of this experiment yourself, go to the first Stroop task on the color insert on the opposite page and time how long it takes you to say the colors of all the words in the "Congruent Color Words" list. When you are done, turn the page and try the depleting condition by timing how long it takes you to say aloud the colors of all the words in the "Incongruent Color Words" list.

How long did these two tasks take you? If you are like most of our participants, reading the congruent list (the nondepleting condition) probably took around sixty seconds, but reading the incongruent list (the depleting condition) was probably three to four times more difficult and more time-consuming.

Somewhat ironically, the difficulty of naming the colors in the mismatched list stems from our skill as readers. For experienced readers, the meaning of the words we read comes to mind very quickly, creating an almost automatic reaction to say the corresponding word rather than the color of the ink. We see the green-colored word "red" and want to say "red!" But that is not what we are supposed to do in this task, so with some effort we suppress our initial response and instead name the color of the ink. You may also have noticed that as you keep at this task, you experience a sort of mental exhaustion resulting from the repeated suppression of your quick automatic responses in favor of the more controlled, effortful (and correct) responses.

After completing either the easy or the hard Stroop task, each participant was given the opportunity to take a multiple-choice quiz about the history of Florida State University. The test included questions such as "When was the school founded?" and "How many times did the football team play

NONDEPLETING CONDITION

Congruent Color Words

RED	BLUE	GREEN	RED	BLUE
GREEN	GREEN	RED	BLUE	GREEN
BLUE	RED	BLUE	GREEN	RED
GREEN	BLUE	RED	RED	BLUE
RED	RED	GREEN	BLUE	GREEN
BLUE	GREEN	BLUE	GREEN	RED
RED	BLUE	GREEN	BLUE	GREEN
BLUE	GREEN	RED	GREEN	RED
GREEN	RED	BLUE	RED	BLUE
BLUE	GREEN	GREEN	BLUE	GREEN
GREEN	RED	BLUE	RED	RED
RED	BLUE	RED	GREEN	BLUE
GREEN	RED	BLUE	RED	GREEN
BLUE	BLUE	RED	GREEN	RED
RED	GREEN	GREEN	BLUE	BLUE

DEPLETING CONDITION

Incongruent Color Words

RED BLUE GREEN RED BLUE
GREEN GREEN RED BLUE GREEN
BLUE RED BLUE GREEN RED
GREEN BLUE RED RED BLUE
RED RED GREEN BLUE GREEN
BLUE GREEN BLUE GREEN RED
RED BLUE GREEN BLUE GREEN
BLUE GREEN RED GREEN RED
GREEN RED BLUE RED BLUE
BLUE GREEN GREEN BLUE GREEN
GREEN RED BLUE RED RED
RED BLUE RED GREEN BLUE
GREEN RED BLUE RED GREEN
BLUE BLUE RED GREEN RED
RED GREEN GREEN BLUE BLUE

in the National Championship game between 1993 and 2001?" In total, the quiz included fifty questions, each with four possible answers, and participants were paid according to their performances. The participants were also told that once they finished answering all the questions, they would be given a bubble sheet so that they could transfer their answers from the quiz to the sheet, recycle the quiz itself, and submit only the bubble sheet for payment.

Imagine that you are a student in the condition with the opportunity to cheat. You have just finished the Stroop task (either the depleting or nondepleting version). You have been answering the quiz questions for the past few minutes, and the time allotted for the quiz is up. You walk up to the experimenter to pick up the bubble sheet so that you can dutifully transfer your answers.

"I'm sorry," the experimenter says, pursing her lips in self-annoyance. "I'm almost out of bubble sheets! I only have one unmarked one, and one that has the answers premarked." She tells you that she did her best to erase the marks on the used bubble sheet but the answers are still slightly visible. Annoyed with herself, she admits that she had hoped to administer one more test today after yours. She next turns to you and asks you a question: "Since you are the first among the last two participants of the day, you can choose which form you would like to use: the clean one or the premarked one."

Of course you realize that taking the premarked bubble sheet would give you an edge if you decided to cheat. Do you take it? Maybe you take the premarked one out of altruism: you want to help the experimenter so that she won't worry so much about it. Maybe you take the premarked one to cheat. Or maybe you think that taking the premarked one would tempt you to cheat, so you reject it because you want to be an honest, upstanding, moral person. Whichever you take, you transfer your answers to that bubble sheet, shred the original quiz, and give the bubble sheet back to the experimenter, who pays you accordingly.

Did the depleted participants recuse themselves from the tempting situation more often, or did they gravitate toward it? As it turned out, they were more likely than nondepleted participants to choose the sheet that tempted them to cheat. As a result of their depletion, they suffered a double whammy: they picked the premarked bubble sheet more frequently, and (as we saw in the previous experiment) they also cheated more when cheating was possible. When we looked at these two ways of cheating combined, we found that we paid the depleted participants 197 percent more than those who were not depleted.

Depletion in Everyday Life

Imagine you're on a protein-and-vegetable diet and you go grocery shopping at the end of the day. You enter the supermarket, vaguely hungry, and detect the smell of warm bread wafting from the bakery. You see fresh pineapple on sale; although you adore it, it is off limits. You wheel your cart to the meat counter to buy some chicken. The crab cakes look good, but they have too many carbohydrates so you pass them by, too. You pick up lettuce and tomatoes for a salad, steeling yourself against the cheesy garlic croutons. You make it to the checkout counter and pay for your goods. You feel very good about yourself and your ability to resist temptation. Then, as you are safely out of the store and on the way to your car, you pass a school bake sale, and a cute little girl offers you a free brownie sample.

Now that you know what you know about depletion, you can predict what your past heroic attempts of resisting temptation may cause you to do: you will most likely give in and take a bite. Having tasted the delicious chocolate melting over your deprived taste buds, you can't possibly walk away. You're dying for more. So you buy enough brownies for a family of eight and end up eating half of them before you even get home.

NOW THINK ABOUT shopping malls. Say you need a new pair of walking shoes. As you make your way from Neiman Marcus to Sears across a vast expanse of gleaming commercial temptation, you see all kinds of things you want but don't necessarily need. There's that new grill set you've been drooling over, that faux-shearling coat for next winter, and the gold necklace for the party you will most likely attend on New Year's Eve. Every enticing item you pass in the window and don't buy is a crushed impulse, slowly whittling away at your reserve of willpower—making it much more likely that later in the day you will fall for temptation.

Being human and susceptible to temptation, we all suffer in this regard. When we make complex decisions throughout the day (and most decisions are more complex and taxing than naming the colors of mismatched words), we repeatedly find ourselves in circumstances that create a tug-of-war between impulse and reason. And when it comes to important decisions (health, marriage, and so on), we experience an even stronger struggle. Ironically, simple, everyday attempts to keep our impulses under control weaken our supply of self-control, thus making us more susceptible to temptation.

NOW THAT YOU know about the effects of depletion, how can you best confront life's many temptations? Here's one approach suggested by my friend Dan Silverman, an economist at the University of Michigan who was facing grave temptation on a daily basis.

Dan and I were colleagues at the Institute for Advanced Study at Princeton. The Institute is a lovely place for lucky researchers who can take a year off to do little else besides think, go for walks in the woods, and eat well. Every day, after we'd spent our mornings pondering life, science, art, and the reason for it all, we enjoyed a delectable lunch: say, duck breast served with polenta and glazed mushroom caps. Each lunch menu was accompanied by a wonderful dessert: ice cream, crème brûlée, New York cheesecake, triple chocolate cake with raspberry-crème filling. It was torturous, particularly for poor Dan, who had a powerful sweet tooth. Being a smart, rational, cholesterolically challenged economist, Dan wanted dessert, but he also understood that eating dessert daily is not advisable.

Dan thought about his problem for a while and concluded that when faced with temptation, a rational person should sometimes succumb. Why? Because by doing so, the rational person can keep him- or herself from becoming too depleted, remaining strong for whatever temptations the future may bring. So for Dan, who was very careful and concerned about future temptations, it was always carpe diem when it came to the daily dessert. And yes, along with Emre Ozdenoren and

Steve Salant, Dan wrote an academic paper justifying this approach.

ON A MORE serious note, these experiments with depletion suggest that, in general, we would be well served to realize that we are continually tempted throughout the day and that our ability to fight this temptation weakens with time and accumulated resistance. If we're really serious about losing weight, we should get rid of temptation by clearing our shelves and refrigerator of all the sugary, salty, fatty, and processed foods and acclimating to the taste of fresh produce. We should do this not only because we know that fried chicken and cake are bad for us but also because we know that exposing ourselves to such temptations throughout the day (and every time we open a cupboard or the refrigerator) makes it more difficult for us to fight off this and other temptations throughout the day.

Understanding depletion also means that (to the extent that we can) we should face the situations that require self-control—a particularly tedious assignment at work, for example—early in the day, before we are too depleted. This, of course, is not easy advice to follow because the commercial forces around us (bars, online shopping, Facebook, YouTube, online computer games, and so on) thrive on both temptation and depletion, which is why they are so successful.

Granted, we cannot avoid being exposed to all threats to our self-control. So is there any hope for us? Here's one suggestion: once we realize that it is very hard to turn away when we face temptation, we can recognize that a better strategy is to walk away from the draw of desire before we are close enough to be snagged by it. Accepting this advice might not be easy, but the reality is that it is much easier to avoid temptation altogether rather than to overcome it when it sits lingering on the kitchen counter. And if we can't quite do that, we can always try to work on our ability to fight temptation—perhaps by counting to a hundred, singing a song, or making an action plan and sticking to it. Doing any of these can help us build our arsenal of tricks for overcoming temptation so that we are better equipped to fight those urges in the future.

FINALLY, I SHOULD point out that sometimes depletion can be beneficial. Occasionally, we may feel that we are too much in control, dealing with too many restrictions, and that we're not sufficiently free to follow our impulses. Perhaps sometimes, we just need to stop being responsible adults and let loose. So here's a tip: next time you really want to let it all hang out and indulge your primal self, try depleting yourself first by writing a long autobiographical essay without using the letters "a" and "n." Then go to a mall, try on different things, but buy nothing. Afterward, with all of this depletion weighing on you, place yourself in the tempting situation of your choice and let 'er rip. Just don't use this trick too often.

AND IF YOU really need a more official-sounding excuse to succumb to temptation from time to time, just use Dan Silverman's theory of rational self-indulgence as the ultimate license.

CHAPTER 5

Why Wearing Fakes Makes Us Cheat More

Let me tell you the story of my debut into the world of fashion. When Jennifer Wideman Green (a friend of mine from graduate school) ended up living in New York City, she met a number of people in the fashion industry. Through her I met Freeda Fawal-Farah, who worked for *Harper's Bazaar*, a gold standard in the fashion industry. A few months later Freeda invited me to give a talk at the magazine, and because it was such an atypical crowd for me, I agreed.

Before I started my talk, Freeda treated me to a quick fashion tutorial as we sipped our lattes in a balcony café overlooking the escalator in the big downtown Manhattan building. Freeda gave me a rundown of the outfits worn by every woman who passed us, including the brands they were wearing and what her clothes and shoes said about her lifestyle. I found her attention to every detail—indeed, the whole fashion analysis—fascinating, the way I imagine expert bird watchers are able to discern minute differences between species.

About thirty minutes later, I found myself on a stage before an auditorium full of fashion mavens. It was a tremendous pleasure to be surrounded by so many attractive and well-dressed women. Each woman was like an exhibit in a museum: her jewelry, her makeup, and, of course, her stunning shoes. Thanks to Freeda's tutorial, I was able to recognize a few of the brands when I looked out into the rows. I could even discern the sense of fashion that inspired each ensemble.

I wasn't sure why those fashionistas wanted me there or what they expected to hear from me. Still, we seemed to have good chemistry. I talked about how people make decisions, how we compare prices when we are trying to figure out how much something is worth, how we compare ourselves to others, and so on. They laughed when I hoped they would, asked thoughtful questions, and offered plenty of their own interesting ideas. When I finished the talk, Valerie Salembier, the publisher of *Harper's Bazaar*, came onstage, hugged and thanked me—and gave me a stylish black Prada overnight bag.

AFTER SAYING OUR good-byes, I left the building with my new Prada bag and headed downtown to my next meeting. I had some time to kill, so I decided to take a walk. As I wandered, I couldn't help thinking about my big black leather bag with its large Prada logo displayed. I debated with myself: should I carry my new bag with the logo facing outward? That way, other people could see and admire it (or maybe just wonder how someone wearing jeans and red sneakers could possibly have procured it). Or should I carry it with the logo facing toward me, so that no one could recognize that it was a Prada? I decided on the latter and turned the bag around.

Even though I was pretty sure that with the logo hidden no one realized it was a Prada bag, and despite the fact that I don't think of myself as someone who cares about fashion, something felt different to me. I was continuously aware of the brand on the bag. I was wearing Prada! And it made me feel different; I stood a little straighter and walked with a bit more swagger. I wondered what would happen if I wore Ferrari underwear. Would I feel more invigorated? More confident? More agile? Faster?

I continued walking and passed through Chinatown, which was bustling with activity, food, smells, and street vendors selling their wares along Canal Street. Not far away, I spotted an attractive young couple in their twenties taking in the scene. A Chinese man approached them. "Handbags, handbags!" he called, tilting his head to indicate the direction of his small shop. At first they didn't react. Then, after a moment or two, the woman asked the Chinese man, "You have Prada?"

The vendor nodded. I watched as she conferred with her partner. He smiled at her, and they followed the man to his stand.

The Prada they were referring to, of course, was not actually Prada. Nor were the $5 "designer" sunglasses on display in his stand really Dolce&Gabbana. And the Armani perfumes displayed over by the street food stands? Fakes too.*

* The market for fake goods, of course, ranges far beyond Chinatown and New York. After gathering momentum for more than forty years, the phenomenon is now a formidable affair. Counterfeiting is illegal almost everywhere on our planet, though the severity of the punishment varies from country to country, as does people's view of the

From Ermine to Armani

Let's pause for a moment and consider the history of wardrobes, thinking specifically about something social scientists call external signaling, which is simply the way we broadcast to others who we are by what we wear. Going back a way, ancient Roman law included a set of regulations called sumptuary laws, which filtered down through the centuries into the laws of nearly all European nations. Among other things, the laws dictated who could wear what, according to their station and class. The laws went into an extraordinary level of detail. For example, in Renaissance England, only the nobility could wear certain kinds of fur, fabrics, laces, decorative beading per square foot, and so on, while those in the gentry could wear decisively less appealing clothing. (The poorest were generally excluded from the law, as there was little point in regulating musty burlap, wool, and hair shirts.)

Some groups were further differentiated so as not to be confused with "respectable" people. For instance, prostitutes had to wear striped hoods to signal their "impurity," and heretics were sometimes forced to don patches decorated with wood bundles to indicate that they could or should be burned at the stake. In a sense, a prostitute going out without her mandatory striped hood was in disguise, like someone wearing a pair of fake Gucci sunglasses. A solid, nonstriped hood sent a false signal of the woman's livelihood and economic status. People who "dressed above their station" were silently, but directly, lying to those around them. Although

morality of buying counterfeits. (See Frederick Balfour, "Fakes!" *BusinessWeek*, February 7, 2005.)

dressing above one's station was not a capital offense, those who broke the law were often hit with fines and other punishments.

What may seem to be an absurd degree of obsessive compulsion on the part of the upper crust was in reality an effort to ensure that people were what they signaled themselves to be; the system was designed to eliminate disorder and confusion. (It clearly had some signaling advantages, though I am not suggesting that we revert back to it.) Although our current sartorial class system is not as rigid as it was in the past, the desire to signal success and individuality is as strong today as ever. The fashionably privileged now wear Armani instead of ermine. And just as Freeda knew that Via Spiga platform heels weren't for everyone, the signals we send are undeniably informative to those around us.

NOW, YOU MIGHT think that the people who buy knockoffs don't actually hurt the fashion manufacturer because many of them would never buy the real thing to start with. But that is where the effect of external signaling comes in. After all, if a bunch of people buy knockoff Burberry scarves for $10, others—the few who can afford the real thing and want to buy it—might not be willing to pay twenty times more for the authentic scarves. If it is the case that when we see a person wearing a signature Burberry plaid or carrying a Louis Vuitton LV-patterned bag, we immediately suspect that it is a fake, then what is the signaling value in buying the authentic version? This perspective means that the people who purchase knockoffs dilute the potency of external signaling and undermine the authenticity of the real product (and its wearer). And that is one reason why fashion retailers and fashionistas care so much about counterfeits.

WHEN THINKING ABOUT my experience with the Prada bag, I wondered whether there were other psychological forces related to fakes that go beyond external signaling. There I was in Chinatown holding my real Prada bag, watching the woman emerge from the shop holding her fake one. Despite the fact that I had neither picked out nor paid for mine, it felt to me that there was a substantial difference between the way I related to my bag and the way she related to hers.

More generally, I started wondering about the relationship between what we wear and how we behave, and it made me think about a concept that social scientists call self-signaling. The basic idea behind self-signaling is that despite what we tend to think, we don't have a very clear notion of who we are. We generally believe that we have a privileged view of our own preferences and character, but in reality we don't know ourselves that well (and definitely not as well as we think we do). Instead, we observe ourselves in the same way we observe and judge the actions of other people—inferring who we are and what we like from our actions.

For example, imagine that you see a beggar on the street. Rather than ignoring him or giving him money, you decide to buy him a sandwich. The action in itself does not define who you are, your morality, or your character, but you interpret the deed as evidence of your compassionate and charitable character. Now, armed with this "new" information, you start believing more intensely in your own benevolence. That's self-signaling at work.

The same principle could also apply to fashion accessories. Carrying a real Prada bag—even if no one else knows it is real—could make us think and act a little differently than if we were carrying a counterfeit one. Which brings us to the questions: Does wearing counterfeit products somehow make us feel less legitimate? Is it possible that accessorizing with fakes might affect us in unexpected and negative ways?

Calling All Chloés

I decided to call Freeda and tell her about my recent interest in high fashion. (I think she was even more surprised than I was.) During our conversation, Freeda promised to convince a fashion designer to lend me some items to use in some experiments. A few weeks later, I received a package from the Chloé label containing twenty handbags and twenty pairs of sunglasses. The statement accompanying the package told me that the handbags were estimated to be worth around $40,000 and the sunglasses around $7,000.*

With those hot commodities in hand, Francesca Gino, Mike Norton (a professor at Harvard University), and I set about testing whether participants who wore fake products would feel and behave differently from those wearing authentic ones. If our participants felt that wearing fakes would broadcast (even to themselves) a less hon-

* The rumor about this shipment quickly traveled around Duke, and I became popular among the fashion-minded crowd.

orable self-image, we wondered whether they might start thinking of themselves as somewhat less honest. And with this tainted self-concept in mind, would they be more likely to continue down the road of dishonesty?

Using the lure of Chloé accessories, we enlisted many female MBA students for our experiment. (We focused on women not because we thought that they were different from men in any moral way—in fact, in all of our previous experiments we did not find any sex-related differences—but because the accessories we had were clearly designed for women.) We wondered whether to use the sunglasses or the handbags in our first experiments, but when we realized that it would have been a bit more difficult to explain why we wanted our participants to walk around the building with handbags, we settled on the sunglasses.

AT THE START of the experiment, we assigned each woman to one of three conditions: authentic, fake, or no information. In the authentic condition, we told participants that they would be donning real Chloé designer sunglasses. In the fake condition, we told them that they would be wearing counterfeit sunglasses that looked identical to those made by Chloé (in actuality all the products we used were the real McCoy). Finally, in the no-information condition, we didn't say anything about the authenticity of the sunglasses.

Once the women donned their sunglasses, we directed them to the hallway, where we asked them to look at different posters and out the windows so that they could later evaluate the quality and experience of looking through their sunglasses. Soon after, we called them into another room for another task. What was the task? You guessed it: while the women were still wearing their sunglasses we gave them our old friend, the matrix task.

Now imagine yourself as a participant in this study. You show up to the lab, and you're randomly assigned to the fake condition. The experimenter informs you that your glasses are counterfeit and instructs you to test them out to see what you think. You're handed a rather real-looking case (the logo is spot-on!), and you pull out the sunglasses, examine them, and slip them on. Once you've put on the specs, you start walking around the hallway, examining different posters and looking out the windows. But while you are doing so, what is going through your head? Do you compare the sunglasses to the pair in your car or the ones you broke the other day? Do you think, "Yeah, these are very convincing. No one would be able to tell they're fake." Maybe you think that the weight doesn't feel right or that the plastic seems cheap. And if you do think about the fakeness of what you are wearing, would it cause you to cheat more on the matrix test? Less? The same amount?

Here's what we found. As usual, lots of people cheated by a few questions. But while "only" 30 percent of the participants in the authentic condition reported solving more matrices than they actually had, 71 percent of those in the fake condition reported solving more matrices than they actually had.

These results gave rise to another interesting question. Did the presumed fakeness of the product make the women cheat more than they naturally would? Or did the genuine Chloé label make them behave more honestly than they would otherwise? In other words, which was more powerful: the negative self-signaling in the fake condition or the positive self-signaling in the authentic condition?

This is why we also had the no-information (control) condition, in which we didn't mention anything about whether the sunglasses were real or fake. How would the no-information condition help us? Let's say that women wearing the fake glasses cheated at the same level as those in the no-information condition. If that were the case, we could conclude that the counterfeit label did not make the women any more dishonest than they were naturally and that the genuine label was causing higher honesty. On the other hand, if we saw that the women wearing the real Chloé sunglasses cheated at the same level as those in the no-information condition (and much less than those in the fake-label condition), we would conclude that the authentic label did not make the women any more honest than they were naturally and that the fake label was causing women to behave less honestly.

As you'll recall, 30 percent of women in the authentic condition and 71 percent of women in the fake-label condition overreported the number of matrices they solved. And in the no-information condition? In that condition 42 percent of the women cheated. The no-information condition was between the two, but it was much closer to the authentic condition (in fact, the two conditions were not statistically different from each other). These results suggest that wearing a genuine product does not increase our honesty (or at least not by much). But once we knowingly put on a counterfeit product, moral constraints loosen to some degree, making it easier for us to take further steps down the path of dishonesty.

The moral of the story? If you, your friend, or someone you are dating wears counterfeit products, be careful! Another act of dishonesty may be closer than you expect.

The "What-the-Hell" Effect

Now let's pause for a minute to think again about what happens when you go on a diet. When you start out, you work hard to stick to the diet's difficult rules: half a grapefruit, a slice of dry multigrain toast, and a poached egg for breakfast; turkey slices on salad with zero-calorie dressing for lunch; baked fish and steamed broccoli for dinner. As we learned in the preceding chapter, "Why We Blow It When We're Tired," you are now honorably and predictably deprived. Then someone puts a slice of cake in front of you. The moment you give in to temptation and take that first bite, your perspective shifts. You tell yourself, "Oh, what the hell, I've broken my diet, so why not have the whole slice—along with that perfectly grilled, mouthwatering cheeseburger with all the trimmings I've been craving all week? I'll start anew tomorrow, or maybe on Monday. And this time I'll really stick to it." In other words, having already tarnished your dieting self-concept, you decide to break your diet completely and make the most of your diet-free self-image (of course you don't take into account that the same thing can happen again tomorrow and the day after, and so on).

To examine this foible in more detail, Francesca, Mike, and I wanted to examine whether failing at one small thing (such as eating one french fry when you're supposedly on a diet) can cause one to abandon the effort altogether.

This time, imagine you're wearing a pair of sunglasses—whether they are authentic Chloé, a fake pair, or a pair of unspecified authenticity. Next, you sit down in front of a computer screen where you're presented with a square divided into two triangles by a diagonal line. The trial starts, and for one second, twenty randomly scattered dots flash within the square (see the diagram below). Then the dots disappear, leaving you with an empty square, the diagonal line, and two response buttons, one marked "more-on-right" and the other marked "more-on-left." Using these two buttons, your task is to indicate whether there were more dots on the right-hand or left-hand side of the diagonal. You do this one hundred times. Sometimes the right-hand side clearly has more dots. Sometimes they are unmistakably concentrated on the left-hand side. Other times it's hard to tell. As you can imagine, you get pretty used to the task, as tedious as it may be, and after a hundred responses the experimenter can tell how accurately you can make these sorts of judgments.

Figure 3: The Dots Task

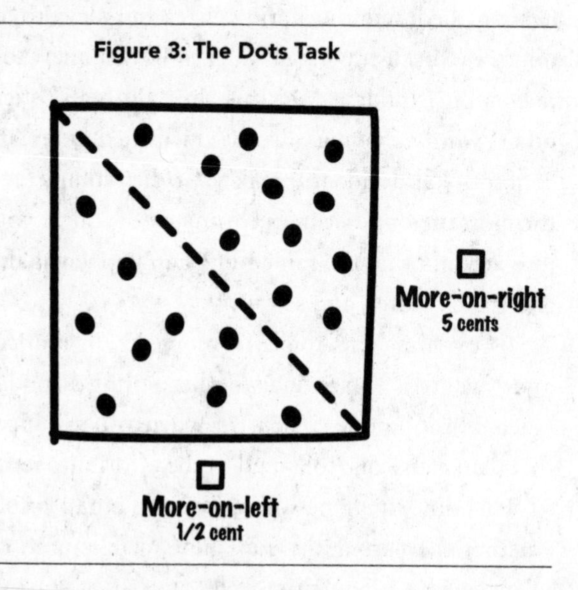

Next, the computer asks you to repeat the same task two hundred more times. Only this time, you will be paid according to your decisions. Here's the key detail: regardless of whether your responses are accurate or not, every time you select the left-hand button, you will receive half a cent and every time you select the right-hand button, you will receive 5 cents (ten times more money).

With this incentive structure, you are occasionally faced with a basic conflict of interest. Every time you see more dots on the right, there is no ethical problem because giving the honest answer (more on the right) is the same response that makes you the most money. But when you see more dots on the left, you have to decide whether to give the accurate honest answer (more on the left), as you were instructed, or to maximize your profit by clicking the more-on-right button. By creating this skewed payment system, we gave the participants an incentive to see reality in a slightly different way and cheat by excessively clicking the more-on-right button. In other words, they were faced with a conflict between producing an accurate answer and maximizing their profit. To cheat or not to cheat, that was the question. And don't forget, you're doing this while still wearing the sunglasses.

As it turned out, our dots task showed the same general results as the matrix task, with lots of people cheating but just by a bit. Interestingly, we also saw that the amount of cheating was especially large for those wearing the fake sunglasses. What's more, the counterfeit wearers cheated more across the board. They cheated more when it was hard to tell which side had more dots, and they cheated more even when it was clear that the correct answer was more on the left (the side with the lower financial reward).

Those were the overall results, but the reason we created the dots task in the first place was to observe how cheating evolves over time in situations where people have many opportunities to act dishonestly. We were interested in whether our participants started the experiment by cheating only occasionally, trying to maintain the belief that they were honest but at the same time benefitting from some occasional cheating. We suspected that this kind of balanced cheating could last for a while but that at some point participants might reach their "honesty threshold." And once they passed that point, they would start thinking, "What the hell, as long as I'm a cheater, I might as well get the most out of it." And from then on, they would cheat much more frequently—or even every chance they got.

The first thing that the results revealed was that the amount of cheating increased as the experiment went on. And as our intuitions had suggested, we also saw that for many people there was a very sharp transition where at some point in the experiment, they suddenly graduated from engaging in a little bit of cheating to cheating at every single opportunity they had. This general pattern of behavior is what we would expect from the what-the-hell effect, and it surfaced in both the authentic and the fake conditions. But the wearers of the fake sunglasses showed a much greater tendency to abandon their moral constraints and cheat at full throttle.*

In terms of the what-the-hell effect, we saw that when it comes to cheating, we behave pretty much the same as we do on diets. Once we start violating our own standards (say, with cheating on diets or for monetary incentives), we

* You might wonder if receiving counterfeits as gifts would have the same effect as choosing a counterfeit product for ourselves. We wondered the same thing and tested this question in another experiment. It turned out that it doesn't matter whether we acquire a counterfeit product by our own choice or not; once we have a fake product, we are more likely to cheat.

are much more likely to abandon further attempts to control our behavior—and from that point on there is a good chance that we will succumb to the temptation to further misbehave.

IT SEEMS, THEN, that the clothes do make the man (or woman) and that wearing knockoffs does have an effect on ethical decisions. As is the case with many findings in social science research, there are ways to use this information for both good and ill. On the negative side, one can imagine how organizations could use this principle to loosen the morality of their employees such that they will find it easier to "fake out" their customers, suppliers, regulators, and competitors and, by doing so, increase the company's revenue at the expense of the other parties. On the positive side, understanding how slippery slopes operate can direct us to pay more attention to early cases of transgression and help us apply the brakes before it's too late.

Up to No Good

Having completed these experiments, Francesca, Mike, and I had evidence that wearing counterfeits colors the way we view ourselves and that once we are painted as cheaters in our own eyes, we start behaving in more dishonest ways. This led us to another question: if wearing counterfeits changes the way we view our own behavior, does it also cause us to be more suspicious of others?

To find out, we asked another group of participants to put on what we told them were either real or counterfeit Chloé sunglasses. Again, they dutifully walked the hall examining different posters and views from the windows. However, when we called them back to the lab, we did not ask them to perform our matrix or dots task. Instead, we asked them to fill out a rather long survey with their sunglasses on. In this survey, we asked a bunch of irrelevant questions (filler questions) that are meant to obscure the real goal of the study. Among the filler questions, we included three sets of questions designed to measure how our respondents interpreted and evaluated the morality of others.

The questions in set A asked participants to estimate the likelihood that people they know might engage in various ethically questionable behaviors. The questions in set B asked them to estimate the likelihood that when people say particular phrases, they are lying. Finally, set C presented participants with two scenarios depicting someone who has the opportunity to behave dishonestly, and they were asked to estimate the likelihood that the person in the scenario would take the opportunity to cheat. Here are the questions from all three sets:

Set A: How likely are people you know to engage in the following behaviors?

- Stand in the express line with too many groceries.
- Try to board a plane before their group number is called.
- Inflate their business expense report.
- Tell their supervisor that progress has been made on a project when none has been made.
- Take home office supplies from work.
- Lie to an insurance company about the value of goods that were damaged.

- Buy a garment, wear it, and return it.
- Lie to their partner about the number of sex partners they have had.

Set B: When the following lines are uttered, how likely is it that they are a lie?

- Sorry I'm late, traffic was terrible.
- My GPA is 4.0.
- It was good meeting you. Let's have lunch sometime.
- Sure, I'll start working on that tonight.
- Yes, John was with me last night.
- I thought I already sent that e-mail out. I am sure I did.

Set C: How likely are these individuals to take the action described?

- Steve is the operations manager of a firm that produces pesticides and fertilizers for lawns and gardens. A certain toxic chemical is going to be banned in a year, and for this reason is extremely cheap now. If Steve buys this chemical and produces and distributes his product fast enough, he will be able to make a very nice profit. Please estimate the likelihood that Steve will sell this chemical while it is still legal.
- Dale is the operations manager of a firm that produces health foods. One of their organic fruit beverages has 109 calories per serving. Dale knows that people are sensitive to crossing the critical threshold of 100 calories. He could decrease the serving size by 10 percent. The label will then say that each serving has 98 calories, and the fine print will say that each bottle contains 2.2 servings. Please estimate the likelihood that Dale will

cut the serving size to avoid crossing the 100-calorie-per-serving threshold.

What were the results? You guessed it. When reflecting on the behavior of people they know (set A), participants in the counterfeit condition judged their acquaintances to be more likely to behave dishonestly than did participants in the authentic condition. They also interpreted the list of common excuses (set B) as more likely to be lies, and judged the actor in the two scenarios (set C) as being more likely to choose the shadier option. In the end, we concluded that counterfeit products not only tend to make us more dishonest; they cause us to view others as less than honest as well.

Fake It Till You Make It

So what can we do with all of these results?

First, let's think about high-fashion companies, which have been up in arms about counterfeits for years. It may be difficult to sympathize with them; you might think that outside their immediate circle, no one should really care about the "woes" of high-end designers who cater to the wealthy. When tempted to buy a fake Prada bag, you might say to yourself, "Well, designer products are too expensive, and it's silly to pay for the real thing." You might say, "I wouldn't consider buying the real product anyway, so the designer isn't really losing any money." Or maybe you would say, "Those fashion companies make so much money that a few people buying fake products won't really make a difference." Whatever rationalizations we come up with—and we are all very good at rationalizing our actions so that they are in line with our selfish motives—it's difficult to find many people who feel that the alarm on the part of high-fashion companies is of grave personal concern.

But our results show that there's another, more insidious story here. High-fashion companies aren't the only ones paying a price for counterfeits. Thanks to self-signaling and the what-the-hell effect, a single act of dishonesty can change a person's behavior from that point onward. What's more, if it's an act of dishonesty that comes with a built-in reminder (think about fake sunglasses with a big "Gucci" stamped on the side), the downstream influence could be long-lived and substantial. Ultimately, this means that we all pay a price for counterfeits in terms of moral currency; "faking it" changes our behavior, our self-image, and the way we view others around us.*

Consider, for example, the fact that academic diplomas hang in many executive suites around the world and decorate even more résumés. A few years ago, *The Wall Street Journal* ran a piece on executives who misrepresent their academic credentials, pointing to top moguls such as Kenneth Keiser, who at the time was the president and COO of PepsiAmericas, Inc. Though Keiser had attended Michigan State University, he never graduated; still, for a long time, he signed off on papers that claimed he had a BA from Michigan State[1] (of course, it is possible that this was just a misunderstanding).

---

* You might wonder if people are aware of the downstream consequences of counterfeits. We tested this too and found that they are unaware of these effects.

Or consider the case of Marilee Jones, who coauthored a popular guidebook called *Less Stress, More Success: A New Approach to Guiding Your Teen Through College Admissions and Beyond*, in which, among other things, she advocated "being yourself" in order to be successful in college admissions and job searches. She was MIT's popular dean of admissions, and for twenty-five years, by all accounts, she did her job very well. There was just one problem: she had added several fictitious degrees to her résumé to land that job in the first place. It was an act of cheating, pure and simple. The irony of her fall from grace was not lost on Jones, who apologized for not "having the courage" to correct the "mistakes" on her fake résumé at any point during her employment. When an extremely popular advocate of "being yourself" is toppled by false credentials, what are the rest of us to think?

If you think about this type of cheating in the context of the "what-the-hell" effect, it might be that fake academic credentials often start innocently enough, perhaps along the lines of "fake it till you make it," but once one such act has been established, it can bring about a looser moral standard and a higher tendency to cheat elsewhere. For example, if an executive holding a fake graduate degree puts constant reminders of his fake degree on his letterhead, business cards, résumé, and website, it's not much of a stretch to imagine that he could also start cheating on expense reports, misrepresenting billable hours, or misusing corporate funds. After all, given the what-the-hell effect, it is possible that one initial act of cheating could increase the executive's general level of self-signaled dishonesty, increasing his fudge factor, which would give rise to further fraud.

THE BOTTOM LINE is that we should not view a single act of dishonesty as just one petty act. We tend to forgive people for their first offense with the idea that it is just the first time and everyone makes mistakes. And although this may be true, we should also realize that the first act of dishonesty might be particularly important in shaping the way a person looks at himself and his actions from that point on—and because of that, the first dishonest act is the most important one to prevent. That is why it is important to cut down on the number of seemingly innocuous singular acts of dishonesty. If we do, society might become more honest and less corrupt over time (for more on this, see chapter 8, "Cheating as an Infection").

(DON'T) STEAL THIS BOOK

Finally, no discussion of designer counterfeits could be complete without mentioning their cousin, illegal downloading. (Imagine experiments similar to the ones on fake sunglasses but using illegally downloaded music or movies.) Allow me to share a story about a time when I learned something interesting about illegal downloads. In this particular case, I was the victim. A few months after *Predictably Irrational* was published, I received the following e-mail:

> *Dear Mr. Ariely,*
>
> *I just finished listening to the illegally downloaded version of your audio book this morning, and I wanted to tell you how much I appreciated it.*

I am a 30-year-old African American male from the inner city of Chicago, and for the last five years, I have been making my living by illegally selling CDs and DVDs. I am the only person in my family who is not in prison or homeless. As the last survivor of a family that represents all that is wrong with America, and as someone breaking the law today, I know it is only a matter of time before I join my family in prison.

Some time ago I got a 9-to-5 job, and was excited at the idea of starting a respectable life, but soon after I started, I quit and went back to my illegal business. This is because of the pain I felt at giving up my illegal business that I had built and nurtured for five years. I owned it, and I couldn't find a job that gave me the same feeling of ownership. Needless to say, I could really relate to your research on ownership.

But something else was equally important in pushing me back to the illegal retail business. In the legal retail store where I worked, people often talked about loyalty and care for their customers, but I don't think they understood what this means. In the illegal industry, loyalty and care are much stronger and more intense than anything I encountered in legal retail. Over the years I have built a network of about 100 people who kindly buy from me. We have become real friends with real connections and developed a level of deep care for one another. These connections and friendships with my clients made it very hard for me to give up the business and their friendship in the process.

I'm happy that I listened to your book.

Elijah

After receiving this e-mail from Elijah, I searched the Internet and found a few free downloadable versions of my audiobook and a few scanned copies of the printed version (which, I have to admit, were high-quality scans, including the front and back covers, all the credits and references, and even the copyright notices, which I particularly appreciated).

No matter where you stand on the "information wants to be free" ideological spectrum, seeing your own work distributed for free without permission makes the whole issue of illegal downloads feel a bit more personal, less abstract, and more complex. On the one hand, I'm very happy that people are reading about my research and hopefully benefitting from it. The more the merrier—after all, that is why I write. On the other hand, I also understand the annoyance of those whose work is being illegally copied and sold. Thankfully I have a day job, but I am certain that if I were to rely on writing as my main source of income, illegal downloads would be less of an intellectual curiosity and much more difficult to swallow.

As for Elijah, I think we made a fair exchange. Sure, he illegally copied my audiobook (and made some money in the process), but I learned something interesting about loyalty and care for customers in the illegal industry and even got an idea for possible future research.

WITH ALL OF this in mind, how can we fight our own moral deterioration, the what-the-hell effect, and the potential of one transgressive act to result in long-term negative effects on our morality? Whether we deal with fashion or other domains of life, it should be clear that one immoral act can make another more likely and that immoral acts in one domain can influence our morality in other domains. That being the case, we should focus on early signs of dishonest behaviors and do our best to cut them down in their budding stages before they reach full bloom.

AND WHAT ABOUT the Prada bag that started this whole research project? I made the only possible rational decision: I gave it to my mother.

CHAPTER 6

Cheating Ourselves

Imagine yourself on a soft, sandy beach. The tide is rolling out, creating a wide swath of wet sand for you to wander along. You're heading to the place where you go from time to time to check out girls. Oh, and you're a feisty blue crab. And in reality, you're going to spar with a few other male crabs to see who will win the favor of the females.

Ahead you see a pretty little thing with cute claws. At the same time, you notice that your competition is quickly closing in. You know that the ideal way to handle the situation is to scare off the other crabs. That way you don't have to get into a fight and risk hurting yourself or, worse, lose your chance to mate. So you have to convince the other crabs that you're bigger and stronger. As you inch closer to your competition, you know you need to emphasize your size. However, if you simply pretend to be larger by standing on your toes and halfheartedly waving your claws around, you will probably give yourself away. What to do?

What you need to do is give yourself a pep talk and start believing that you are, in fact, bigger and tougher than you really are. "Knowing" you're the biggest crab on the beach, you stand as high as you can on your hind legs and spread your claws as far and high above you as possible (antlers, peacock tails, and general puffing up help other male creatures do the same thing). Believing in your own fabrication means that you will not flinch. And your (exaggerated) self-confidence might cow your opponents.

NOW BACK TO us. As humans, we have slightly more sophisticated means of puffing ourselves up than our animal counterparts. We have the ability to lie—not just to others but also to ourselves. Self-deception is a useful strategy for believing the stories we tell, and if we are successful, it becomes less likely that we will flinch and accidentally signal that we're anything other than what we pretend to be. I'm hardly endorsing lying as a means of attaining a partner, a job, or anything else. But in this chapter, we'll look at the ways we succeed in fooling ourselves as we try to fool others.

Of course, we can't instantly believe every one of our lies. For instance, let's say you're a guy at a speed-dating event and you're trying to impress an attractive woman. A wild idea enters your mind: you tell her that you have a pilot's license. Even if you sold her this story, it's unlikely you will convince yourself that you do, in fact, have such a license and start suggesting to the pilots on your next flight how to improve their landings. On the other hand, let's say you go out running with your buddy and you get into a discussion about best running times. You tell your friend that you've run a mile in under seven minutes, when in reality your best time was a tiny bit over seven minutes. A few days later, you tell someone else the same thing. After repeating this slightly exaggerated claim over and over, you could eventually forget that you hadn't actually broken the seven-minute mark. You may come to believe it to such a degree that you might even be willing to bet money on it.

ALLOW ME TO tell you a story of a time when I embraced my own deception. In the summer of 1989—about two years after I left the hospital—my friend Ken and I decided to fly from New York to London to see another friend. We bought the cheapest flight to London, which happened to be on Air India. When the taxi dropped us off at the airport, we were dismayed to see a line of people trailing all the way out of the terminal. Thinking fast, Ken came up with an idea: "Why don't we put you in a wheelchair?" I thought about his suggestion. Not only would I be more comfortable, but we could also get through much faster. (Truthfully speaking, it is difficult for me to stand for a prolonged time because the circulation in my legs is far from good. But I don't need a wheelchair.)

We were both convinced that it was a good plan, so Ken jumped out of the cab and returned with the wheelchair. We breezed through check-in and, with an extra two hours to kill, we enjoyed coffee and a sandwich. But then I needed to use the restroom. So Ken pushed me in the wheelchair to the nearest bathroom, which unfortunately was not designed to accommodate a wheelchair. I maintained my role, though; we got the wheelchair as close to the toilet as possible and I tried to hit the mark from a distance, with limited success.

Once we made it through the bathroom challenge, it was time to board the plane. Our seats were in row 30, and as we neared the entrance to the plane, I realized that the wheelchair was going to be too wide for the aisle. So we did what my new role dictated: I left the wheelchair at the entrance of the plane, grabbed on to Ken's shoulders, and he hauled me to our seats.

As I sat waiting for the flight to take off, I was annoyed that the bathroom in the airport wasn't handicap-accessible and that the airline hadn't provided me with a narrower wheelchair to get to my seat. My irritation increased when I realized that I shouldn't drink anything on the six-hour flight because there would be no way for me to keep up the act and use the bathroom. The next difficulty arose when we landed in London. Once again, Ken had to carry me to the entrance of the plane, and when the airline didn't have a wheelchair waiting for us, we had to wait.

This little adventure made me appreciate the daily irritations of handicapped people in general. In fact, I was so annoyed that I decided to go and complain to the head of Air India in London. Once we got the wheelchair, Ken rolled me to Air India's office, and with an overblown air of indignation I described each difficulty and humiliation and reprimanded the regional head of Air India for the airline's lack of concern for disabled people everywhere. Of course he apologized profusely, and after that we rolled away.

The odd thing is that throughout the process I knew I could walk, but I adopted my role so quickly and thoroughly that my self-righteousness felt as real as if I had a legitimate reason to be upset. Then after all that, we got to the baggage claim, where I simply picked up my backpack and walked away unhampered, like Keyser Söze in the film *The Usual Suspects*.

TO MORE SERIOUSLY examine self-deception, Zoë Chance (a postdoc at Yale), Mike Norton, Francesca Gino, and I set out to learn more about how and when we deceive ourselves into believing our own lies and whether there are ways to prevent ourselves from doing so.

In the first phase of our exploration, participants took an eight-question IQ-like test (one of the questions, for example, was this: "What is the number that is one half of one quarter of one tenth of 400?"). After they finished taking the quiz, participants in the control group handed their answers over to the experimenter who checked their responses. This allowed us to establish the average performance on the test.*

In the condition where cheating was possible, participants had an answer key at the bottom of the page. They were told that the answer key was there so that they could score how well they did on the test and also to help them estimate in general how good they were at answering these types of questions. However, they were told to answer the questions first and only then use the key for verification. After answering all the questions, participants checked their own answers and reported their own performance.

What did the results from phase one of the study show? As we expected, the group that had the opportunity to "check their answers" scored a few points higher on average, which suggested that they had used the answer key not only to score themselves but also to improve their performance. As was the case with all of our other experiments, we found that people cheat when they have a chance to do so, but not by a whole lot.

---

* We used this type of SAT-like question instead of our standard matrices because we expected that such questions would lead more naturally to the feeling of "I knew it all along" and to self-deception.

Helping Myself to a Higher MENSA Score

The inspiration for this experimental setup came from one of those complimentary magazines that you find in seat-back pockets on airplanes. On one particular flight, I was flipping through a magazine and discovered a MENSA quiz (questions that are supposed to measure intelligence). Since I am rather competitive, I naturally had to try it. The directions said that the answers were in the back of the magazine. After I answered the first question, I flipped to the back to see if I was correct, and lo and behold, I was. But as I continued with the quiz, I also noticed that as I was checking the answer to the question I just finished solving, my eyes strayed just a bit to the next answer. Having glanced at the answer to the next question, I found the next problem to be much easier. At the end of the quiz, I was able to correctly solve most of the questions, which made it easier for me to believe that I was some sort of genius. But then I had to wonder whether my score was that high because I was supersmart or because I had seen the answers out of the corner of my eye (my inclination was, of course, to attribute it to my own intelligence).

The same basic process can take place in any test in which the answers are available on another page or are written upside down, as they often are in magazines and SAT study guides. We often use the answers when we practice taking tests to convince ourselves that we're smart or, if we get an answer wrong, that we've made a silly mistake that we would never make during a real exam. Either way, we come away with an inflated idea of how bright we actually are—and that's something we're generally happy to accept.

THE RESULTS FROM phase one of our experiments showed that participants tended to look ahead at the answers as a way to improve their score. But this finding did not tell us whether they engaged in straight-up old-fashioned cheating or if they were actually deceiving themselves. In other words, we didn't yet know if the participants knew they were cheating or if they convinced themselves that they legitimately knew the correct answers all along. To figure this out, we added another component to our next experiment.

Imagine that you are taking part in an experiment similar to the previous one. You took the eight-question quiz and answered four questions correctly (50 percent), but thanks to the answers at the bottom of the page, you claimed that you had solved six correctly (75 percent). Now, do you think that your actual ability is in the 50 percent range, or do you think it is in the 75 percent range? On the one hand, you may be aware that you used the answer key to inflate your score, and realize that your real ability is closer to the 50 percent mark. On the other hand, knowing that you were paid as if you really had solved six problems, you might be able to convince yourself that your ability to solve such questions is in reality closer to the 75 percent level.

This is where phase two of the experiment comes in. After finishing the math quiz, the experimenter asks you to predict how well you will do on the next test, in which you will be asked to answer a hundred questions of a similar nature. This time, it's clear that there are not going to be any answers at the bottom of the page (and therefore no chance to consult the key). What do you predict your performance will be on the next quiz? Will it be based on your real ability in the first phase (50 percent), or will it be based on your exaggerated ability (75 percent)? Here is the logic: if you are aware that you used the answer key in the previous test to artificially inflate your score, you would predict that you would correctly solve the same proportion of questions as you solved unassisted in the first test (four out of eight, or around 50 percent). But let's say you started believing that you really did answer six questions correctly on your own and not because you looked at the answers. Now you might predict that in this next test, too, you would correctly solve a much larger percentage of the questions (closer to 75 percent). In truth, of course, you can solve only about half of the questions correctly, but your self-deception may puff you up, crablike, and increase your confidence in your ability.

The results showed that participants experienced the latter sort of self-puffery. The predictions of how well they would perform on the second phase of the test showed that participants not only used the answer key in the first phase to exaggerate their score, but had very quickly convinced themselves that they truly earned that score. Basically, those who had a chance to check their answers in the first phase (and cheated) started believing that their exaggerated performance was a reflection of their true skill.

But what would happen if we paid participants to predict their score accurately in the second phase? With money on the line, maybe our participants wouldn't so patently ignore the fact that in phase one they had used the answer key to improve their scores. To that end, we repeated the same experiment with a new group of participants, this time offering them up to $20 if they correctly predicted their performance on the second test. Even with a financial incentive to be accurate, they still tended to take full credit for their scores and overestimate their abilities. Despite having a strong motivation to be accurate, self-deception ruled the day.

I KNEW IT ALL ALONG

I give a considerable number of lectures about my research to different groups, from academics to industry types. When I started giving talks, I would often describe an experiment, the results, and finally what I thought we could learn from it. But I often found that some people were rather unsurprised by the results and were eager to tell me so. I found this puzzling because, as the person who actually carried out the research, I'd often been surprised by the outcomes myself. I wondered, were the people in the audience really that insightful? How did they know the results sooner than I did? Or was it just an ex post facto feeling of intuition?

Eventually I discovered a way to combat this "I knew it all along" feeling. I started asking the audience to predict the results of the experiments. After I finished describing the setup and what we measured, I gave them a few seconds to think about it. Then I would ask them to vote on the outcome or write their prediction down. Only once they committed to their answer would I provide the results. The good news is that this approach works. Using this ask-first method, I rarely receive the "I knew it all along" response.

In honor of our natural tendency to convince ourselves that we knew the right answers all along, I call my research center at Duke University "The Center for Advanced Hindsight."

Our Love of Exaggeration

Once upon a time—back in the early 1990s—the acclaimed movie director Stanley Kubrick began hearing stories through his assistant about a man who was pretending to be him. The man-who-would-be-Kubrick (whose real name was Alan Conway and who looked nothing like the dark-bearded director) went around London telling people who he famously was(n't). Since the real Stanley Kubrick was a very private person who shunned the paparazzi, not many people had any idea of what he looked like. So a lot of gullible people, thrilled to "know" the famous director personally, eagerly took Conway's bait. Warner Bros., which financed and distributed Kubrick's films, began calling Kubrick's office practically every day with new complaints from people who could not understand why "Stanley" would not get back to them. After all, they had treated him to drinks and dinner and paid for his cab, and he had promised them a part in his next film!

One day, Frank Rich (the former theater critic and op-ed columnist of *The New York Times*) was having dinner in a London restaurant with his wife and another couple. As it happened, the Kubrick imitator was sitting at a nearby table with a knighted MP and a few other young men, regaling them with stories of his moviemaking marvels. When the imposter saw Rich at the next table, he walked over to him and told the critic that he was inclined to sue the *Times* for having called him "creatively dormant." Rich, excited to meet the reclusive "Kubrick," asked for an interview. Conway told Rich to call him, gave Rich his home phone number, and . . . disappeared.

Very shortly after this encounter, things began to unravel for Conway as it dawned on Rich and others that they'd been conned. Eventually the truth came out when Conway began selling his story to journalists. He claimed to be a recovering victim of a mental disorder ("It was uncanny. Kubrick just took me over. I really did believe I was him!"). In the end Conway died a penniless alcoholic, just four months before Kubrick.*

Although this story is rather extreme, Conway may well have believed that he was Kubrick when he was parading around in disguise, which raises the question of whether some of us are more prone to believe our own fibs than others. To examine this possibility, we set up an experiment that repeated the basic self-deception task, but this time we also measured participants' general tendency to turn a blind eye to their own failures. To measure this tendency, we asked participants to agree or disagree with a few statements, such as "My first impressions of people are usually right" and "I never cover up my mistakes." We wanted to see whether people who answered "yes" to more of these questions also had a higher tendency for self-deception in our experiment.

Just as before, we saw that those in the answer-key condition cheated and got higher scores. Again, they predicted that they would correctly answer more questions in the following test. And once more, they lost money because they exaggerated their scores and overpredicted their ability. And what about those who answered "yes" to more of the statements about their own propensities? There were many of them, and they were the ones who predicted that they would do best on our second-phase test.

---

* The story was written up by Kubrick's assistant, Anthony Frewin, in *Stop Smiling* magazine, and it was the basis of the film *Colour Me Kubrick*, starring John Malkovich as Conway.

HEROIC VETERANS?

In 1959, America's "last surviving Civil War veteran," Walter Williams, died. He was given a princely funeral, including a parade that tens of thousands gathered to see, and an official week of mourning. Many years later, however, a journalist named William Marvel discovered that Williams had been only five years old when the war began, which meant he wouldn't have been old enough at any point to serve in the military in any capacity. It gets worse, though. The title that Walter Williams bore falsely to the grave had been passed to him from a man named John Salling, who, as Marvel discovered, had also falsely called himself the oldest Civil War veteran. In fact, Marvel claims that the last dozen of so-called oldest Civil War veterans were all phony.

There are countless other stories like these, even in recent wars, where one might think it would be more difficult to make up and sustain such claims. In one example, Sergeant Thomas Larez received multiple gunshot wounds fighting the Taliban in Afghanistan while helping an injured soldier to safety. Not only did he save his friend's life, but he rallied from his own wounds and killed seven Taliban fighters. So went the reporting of Larez's exploits aired by a Dallas news channel, which later had to run a retraction when it turned out that although Larez was indeed a marine, he had never been anywhere near Afghanistan—the entire story was a lie.

Journalists often uncover such false claims. But once in a while, it's the journalist who's the fibber. With teary eyes and a shaky voice, the longtime journalist Dan Rather described his own career in the marines, even though he had never made it out of basic training. Apparently, he must have believed that his involvement was far more significant than it actually was.[1]

THERE ARE PROBABLY many reasons why people exaggerate their service records. But the frequency of stories about people lying on their résumés, diplomas, and personal histories brings up a few interesting questions: Could it be that when we lie publicly, the recorded lie acts as an achievement marker that "reminds" us of our false achievement and helps cement the fiction into the fabric of our lives? So if a trophy, ribbon, or certificate recognizes something that we never achieved, would the achievement marker help us hold on to false beliefs about our own ability? Would such certificates increase our capacity for self-deception?

BEFORE I TELL you about our experiments on this question I should point out that I proudly hang two diplomas on my office wall. One is an "MIT Bachelor of Science in Charm," and the other is a "PhD in Charm," also from MIT. I was awarded these diplomas by the Charm School, which is an activity that takes place at MIT during the cold and miserable month of January. To fulfill the requirements, I had to take many classes in ballroom dancing, poetry, tie tying, and other such cotillion-inspired skills. And in truth, the longer I have the certificates on my office wall, the more I believe that I am indeed quite charming.

WE TESTED THE effects of certificates by giving our participants a chance to cheat on our first math test (by giving them access to the answer key). After they exaggerated their performance, we gave some of them a certificate emphasizing their (false) achievement on that test. We even wrote their name and score on the certificate and printed it on nice, official-looking paper. The other participants did not receive a certificate. Would the achievement markers raise the participants' confidence in their overstated performance, which in reality was partially based on consulting the answer key? Would it make them believe that their score was, indeed, a true reflection of their ability?

As it turns out, I am not alone in being influenced by diplomas hanging on the wall. The participants who received a certificate predicted that they would correctly answer more questions on the second test. It looks as though having a reminder of a "job well done" makes it easier for us to think that our achievements are all our own, regardless of how well the job was actually done.

THE NINETEENTH-CENTURY NOVELIST Jane Austen provided a fantastic example of the way our own selfish interests, together with the help of others around us, can get us to believe that our selfishness is really a mark of charity and generosity. In *Sense and Sensibility* there is a revealing scene in which John, the first and only son and legal heir, considers what, exactly, is involved in a promise he made to his father. At his father's deathbed, John promises the old man to take care of his very kind but poor stepmother and three half sisters. Of his own accord, he decides to give the women £3,000, a mere fraction of his inheritance, which would take care of them nicely. After all, he genially reasons, "he could spare so considerable a sum with little inconvenience."

Despite the satisfaction John gets from this idea and the ease with which the gift can be given, his clever and selfish wife convinces him—without much difficulty and with a great deal of specious reasoning—that any money he gives his stepfamily will leave him, his wife, and their son "impoverished to a most dreadful degree." Like a wicked witch from a fairy tale, she argues that his father must have been light-headed. After all, the old man was minutes from death when he made the request. She then harps on the stepmother's selfishness. How can John's stepmother and half sisters think they deserve any money? How can he, her husband, squander his father's fortune by providing for his greedy stepmom and sisters? The son, brainwashed, concludes that "It would be absolutely unnecessary, if not highly indecorous, to do more for the widow and his father's three daughters . . ." Et voilà! Conscience appeased, avarice rationalized, fortune untouched.

SELF-DECEPTION IN SPORTS

All players know that steroid use is against the rules and that if they are ever discovered using them it will tarnish their records as well as the sport. Yet the desire to beat new (steroid-fueled) records and to win media attention and fan adoration drives many athletes to cheat by doping. The problem is everywhere and in every sport. There was Floyd Landis, who was stripped of his Tour de France victory because of steroid use in 2006. The University of Waterloo in Canada suspended its entire football team for a year when eight players tested positive for anabolic steroids. A Bulgarian soccer coach was banned from the sport for four years for

giving players steroids before a match in 2010. And yet we can only wonder what steroid users think as they win a match or while receiving a medal. Do they recognize that their praise is undeserved, or do they truly believe that their performance is a pure tribute to their own skill?

Then, of course, there's baseball. Would Mark McGwire hold so many records if not for steroid use? Did he believe his achievement was owing to his own skill? After admitting to steroid use, McGwire stated, "I'm sure people will wonder if I could have hit all those home runs had I never taken steroids. I had good years when I didn't take any, and I had bad years when I didn't take any. I had good years when I took steroids, and I had bad years when I took steroids. But no matter what, I shouldn't have done it and for that I'm truly sorry."[2]

Sorry he may be, but in the end neither his fans nor McGwire himself can know exactly how good he really is.

AS YOU CAN see, people tend to believe their own exaggerated stories. Is it possible to stop or at least decrease this behavior? Since offering money to people to judge their performance more accurately did not seem to eliminate self-deception, we decided to intervene beforehand, right at the moment people were tempted with the opportunity to cheat. (This approach is related to our use of the Ten Commandments in chapter 2, "Fun with the Fudge Factor.") Since our participants were clearly able to ignore the effect that the answer key had on their scores, we wondered what would happen if we made the fact that they were relying on the answer key more obvious at the moment that they were using it. If using the answer key to boost their scores was blatantly obvious, would they be less able to convince themselves that they had known the correct answer all along?

In our initial (paper-based) experiments, it was not possible to figure out exactly when our participants' eyes wandered to the answer key and the level to which they were aware of the help that they got from the written answers. So in our next experiment, we had our participants take a computerized version of the same test. This time the answer key at the bottom of the screen was initially hidden from sight. To reveal the answers, participants had to move the cursor to the bottom of the screen, and when the cursor was moved away, the answer key was hidden again. That way the participants were forced to think about exactly when and for how long they used the answer key, and they could not as easily ignore such a clear and deliberate action.

Although almost all of the participants consulted the answer key at least once, we found that this time around (in contrast to the paper-based tests) they did not overestimate their performance in the second test. Despite the fact that they still cheated, consciously deciding to use the answer key—rather than merely glancing at the bottom of the page—eliminated their self-deceptive tendencies. It seems, then, that when we are made blatantly aware of the ways we cheat, we become far less able to take unwarranted credit for our performance.

Self-deception and Self-help

So where do we stand on self-deception? Should we maintain it? Eliminate it? I suspect that self-deception is similar to its cousins, overconfidence and optimism, and as with these other biases, it has both benefits and disadvantages. On the positive side, an unjustifiably elevated belief in ourselves can increase our general well-being by helping us cope with stress; it can increase our persistence while doing difficult or tedious tasks; and it can get us to try new and different experiences.

We persist in deceiving ourselves in part to maintain a positive self-image. We gloss over our failures, highlight our successes (even when they're not entirely our own), and love to blame other people and outside circumstances when our failures are undeniable. Like our friend the crab, we can use self-deception to boost our confidence when we might not otherwise feel bold. Positioning ourselves on the basis of our finer points can help us snag a date, finish a big project, or land a job. (I am not suggesting that you puff up your résumé, of course, but a little extra confidence can often work in our favor.)

On the negative side, to the extent that an overly optimistic view of ourselves can form the basis of our actions, we may wrongly assume that things will turn out for the best and as a consequence not actively make the best decisions. Self-deception can also cause us to "enhance" our life stories with, say, a degree from a prestigious university, which can lead us to suffer a great deal when the truth is ultimately revealed. And, of course, there is the general cost of deception. When we and those around us are dishonest, we start suspecting everyone, and without trust our lives become more difficult in almost every way.

As in other aspects of life, here too the balance lies between happiness (partially driven by self-deception) and optimal decisions for the future (and a more realistic view of ourselves). Sure, it is exciting to be bright-eyed, with hopes for a wonderful future—but in the case of self-deception, our exaggerated beliefs can devastate us when reality comes crashing in.

Some Upsides of Lying

When we lie for another person's benefit, we call it a "white lie." When we tell a white lie, we're expanding the fudge factor, but we're not doing it for selfish reasons. For example, consider the importance of insincere compliments. We all know the gold standard of white lies, in which a woman who is less than svelte puts on a slinky new dress and asks her husband, "Do I look fat in this?" The man does a quick cost-benefit analysis; he sees his whole life pass before his eyes if he answers with the brutal truth. So he tells her, "Darling, you look beautiful." Another evening (marriage) saved.

Sometimes white lies are just social niceties, but other times they can work wonders to help people get through the most difficult of circumstances, as I learned as an eighteen-year-old burn victim.

After an accident that nearly killed me, I found myself in the hospital with third-degree burns covering over 70 percent of my body. From the beginning, the doctors and the nurses kept telling me, "Everything will be okay." And I wanted to believe them. To my young mind, "Everything will be okay" meant that the scars from my burns and many, many skin transplants would eventually fade and go away, just as when someone burns himself while making popcorn or roasting marshmallows over a campfire.

One day toward the end of my first year in the hospital, the occupational therapist said she wanted to introduce me to a recovered burn victim who'd suffered a similar fate a decade earlier. She wanted to demonstrate to me that it was possible for me to go out into the world and do things that I used to do—basically, that everything would be okay. But when the visitor came in, I was horrified. The man was badly scarred—so badly that he looked deformed. He was able to move his hands and use them in all kinds of creative ways, but they were barely functional. This image was far from the way I imagined my own recovery, my ability to function, and the way I would look once I left the hospital. After this meeting I became deeply depressed, realizing that my scars and functionality would be much worse than I had imagined up to that point.

The doctors and nurses told me other well-meaning lies about what kind of pain to expect. During one unbearably long operation on my hands, the doctors inserted long needles from the tips of my fingers through the joints in order to hold my fingers straight so that the skin could heal properly. At the top of each needle they placed a cork so that I couldn't unintentionally scratch myself or poke my eyes. After a couple of months of living with this unearthly contraption, I found that it would be removed in the clinic—not under anesthesia. That worried me a lot, because I imagined that the pain would be pretty awful. But the nurses said, "Oh, don't worry. This is a simple procedure and it's not even painful." For the next few weeks I felt much less worried about the procedure.

When the time came to withdraw the needles, one nurse held my elbow and the other slowly pulled out each needle with pliers. Of course, the pain was excruciating and lasted for days—very much in contrast to how they described the procedure. Still, in hindsight, I was very glad they had lied to me. If they had told me the truth about what to expect, I would have spent the weeks before the extraction anticipating the procedure in misery, dread, and stress—which in turn might have compromised my much-needed immune system. So in the end, I came to believe that there are certain circumstances in which white lies are justified.

CHAPTER 7

Creativity and Dishonesty

We Are All Storytellers

Facts are for people who lack the imagination
to create their own truth.
—ANONYMOUS

Once upon a time, two researchers named Richard Nisbett (a professor at the University of Michigan) and Tim Wilson (a professor at the University of Virginia) set up camp at their local mall and laid out four pairs of nylon stockings across a table. They then asked female passersby which of the four they liked best. The women voted, and, by and large, they preferred the pair on the far right. Why? Some said they liked the material more. Some said they liked the texture or the color. Others felt that the quality was the highest. This preference was interesting, considering that all four pairs of stockings were identical. (Nisbett and Wilson later repeated the experiment with nightgowns, and found the same results.)

When Nisbett and Wilson questioned each participant about the rationale behind her choice, not one cited the placement of the stockings on the table. Even when the researchers told the women that all the stockings were identical and that there was simply a preference for the right-hand pair, the women "denied it, usually with a worried glance at the interviewer suggesting that they felt either that they had misunderstood the question or were dealing with a madman."

The moral of this story? We may not always know exactly why we do what we do, choose what we choose, or feel what we feel. But the obscurity of our real motivations doesn't stop us from creating perfectly logical-sounding reasons for our actions, decisions, and feelings.

YOU CAN THANK (or perhaps blame) the left side of your brain for this incredible ability to confabulate stories. As the cognitive neuroscientist Michael Gazzaniga (a professor at the University of California, Santa Barbara) puts it, our left brain is "the interpreter," the half that spins a narrative from our experiences.

Gazzaniga came to this conclusion after many years of research with split-brain patients, a rare group whose corpora callosa—the largest bundle of nerves connecting our brain's two hemispheres—had been cut (usually as a way to reduce epileptic seizures). Interestingly, this brain abnormality means that these individuals can be presented with a stimulus to one half of the brain without the other half having any awareness of it.

Working with a female patient who had a severed corpus callosum, Gazzaniga wanted to find out what happens when you ask the right side of the brain to do something and then ask the left side (which has no information about what is going on in the right side) to provide a reason for that action. Using a device that showed written instructions to the patient's right hemisphere, Gazzaniga instructed the right side of the patient's brain to make her laugh by flashing the word "laugh." As soon as the woman complied, he asked her why she had laughed. The woman had no idea why she laughed, but rather than responding with "I don't know," she made up a story. "You guys come up and test us every month. What a way to make a living!" she said. Apparently she had decided that cognitive neuroscientists were pretty amusing.

This anecdote illustrates an extreme case of a tendency we all have. We want explanations for why we behave as we do and for the ways the world around us functions. Even when our feeble explanations have little to do with reality. We're storytelling creatures by nature, and we tell ourselves story after story until we come up with an explanation that we like and that sounds reasonable enough to believe. And when the story portrays us in a more glowing and positive light, so much the better.

Cheating Myself

In a commencement speech at Cal Tech in 1974, the physicist Richard Feynman told graduates, "The first principle is that you must not fool yourself—and you are the easiest person to fool." As we have seen so far, we human beings are torn by a fundamental conflict—our deeply ingrained propensity to lie to ourselves and to others, and the desire to think of ourselves as good and honest people. So we justify our dishonesty by telling ourselves stories about why our actions are acceptable and sometimes even admirable. Indeed, we're pretty skilled at pulling the wool over our own eyes.

Before we examine in more detail what makes us so good at weaving self-glorifying tales, allow me to tell you a little story about how I once (very happily) cheated myself. Quite a few years ago (when I was thirty), I decided that I needed to trade in my motorcycle for a car. I was trying to decide which car would be the perfect one for me. The Internet was just starting to boom with what I'll politely call "decision aids," and to my delight I found a website that provided advice for purchasing cars. The website was based on an interview procedure, and it presented me with a lot of questions that ranged from preferences for price and safety to what kind of headlights and brakes I wanted.

It took about twenty minutes to answer all the questions. Each time I completed a page of answers, I could see the progress bar indicating that I was that much closer to discovering my personalized dream car. I finished the final page of questions and eagerly clicked the "Submit" button. In just a few seconds I got my answer. What was the perfect car for me? According to this finely tuned website, the car for me was . . . drum roll, please . . . a Ford Taurus!

I confess that I did not know much about cars. In fact, I know very little about cars. But I certainly knew that I did not want a Ford Taurus.*

I'm not sure what you would do in such a situation, but I did what any creative person might do: I went back into the program and "fixed" my previous answers. From time to time

* I have nothing against the Ford Taurus, which I am sure is a fine automobile; it just wasn't as exciting a car as I'd imagined myself driving.

I checked to see how different answers translated into different car recommendations. I kept this up until the program was kind enough to recommend a small convertible—surely the right car for me. I followed that sage advice, and that's how I became the proud owner of a convertible (which, by the way, has served me loyally for many years).

This experience taught me that sometimes (perhaps often) we don't make choices based on our explicit preferences. Instead, we have a gut feeling about what we want, and we go through a process of mental gymnastics, applying all kinds of justifications to manipulate the criteria. That way, we can get what we really want, but at the same time keep up the appearance—to ourselves and to others—that we are acting in accordance with our rational and well-reasoned preferences.

Coin Logic

If we accept that we frequently make decisions in this way, perhaps we can make the process of rationalization more efficient and less time-consuming. Here's how: Imagine that you're choosing between two digital cameras. Camera A has a nice zoom and a hefty battery, while camera B is lighter and has a snazzier shape. You're not sure which one to get. You think that camera A is better quality but camera B will make you happier because you like how it looks. What should you do? Here is my advice: Pull a quarter out of your pocket and say to yourself, "Camera A is heads, camera B is tails." Then toss the coin. If the coin comes up heads and camera A is the one you wanted, good for you, go buy it. But if you're not happy with the outcome, start the process again, saying to yourself, "The next toss is for real." Do this until the coin gives you tails. You'll not only get camera B, which you really wanted all along, but you can justify your decision because you only followed the "advice" of the coin. (You could also replace the coin with your friends and consult them until one of them gives you the advice you want.)

Perhaps that was the real function of the car recommendation software I used to get my convertible. Maybe it was designed not only to help me make a better decision but to create a process that would allow me to justify the choice I really wanted to make. If that is the case, I think it would be useful to develop many more of these handy applications for many other areas of life.

The Liar's Brain

Most of us think that some people are especially good (or bad) at deception. If this is indeed the case, what characteristics distinguish them? A team of researchers led by Yaling Yang (a postdoc at the University of California, Los Angeles) tried to find out the answer to this question by studying pathological liars—that is, people who lie compulsively and indiscriminately.

To find participants for their study, Yang and her colleagues went to a Los Angeles temporary employment agency. They figured that at least a few of those who were without permanent employment would have had difficulty holding a job because they were pathological liars. (Obviously, this doesn't apply to all temps.)

The researchers then gave 108 job seekers a battery of psychological tests and conducted several one-on-one interviews with them, their coworkers, and their family members in order to identify major discrepancies that might reveal the pathological liars. In this group, they found twelve people who had pervasive inconsistencies in the stories they told about their work, schooling, crimes committed, and family background. They were also the same individuals who frequently engaged in malingering, or pretending that they were sick in order to get sickness benefits.

Next, the team put the twelve pathological liars—plus twenty-one people who were not pathological liars and were in the same pool of job seekers (the control group)—through a brain scanner to explore each person's brain structure. The researchers focused on the prefrontal cortex, a part of the brain that sits just behind our foreheads and is considered to be in charge of higher-order thinking, such as planning our daily schedule and deciding how to deal with temptations around us. It's also the part of the brain that we depend on for our moral judgments and decision making. In short, it's a kind of control tower for thinking, reasoning, and morality.

In general, there are two types of matter that fill our brains: gray and white. Gray matter is just another name for the collections of neurons that make up the bulk of our brains, the stuff that powers our thinking. White matter is the wiring that connects those brain cells. We all have both gray and white matter, but Yang and her collaborators were particularly interested in the relative amounts of the two types in the participants' prefrontal cortices. They found that pathological liars had 14 percent less gray matter than the control group, a common finding for many psychologically impaired individuals. What could this mean? One possibility is that since the pathological liars had fewer brain cells (the gray matter) fueling their prefrontal cortex (an area crucial to distinguishing between right and wrong), they find it harder to take morality into account, making it easier for them to lie.

But that's not all. You might wonder about the extra space that pathological liars must have in their skulls since they have so much less gray matter. Yang and her colleagues also found that pathological liars had 22 to 26 percent more white matter in the prefrontal cortex than non–pathological liars. With more white matter (remember, this is what links the gray matter), pathological liars are likely able to make more connections between different memories and ideas, and this increased connectivity and access to the world of associations stored in their gray matter might be the secret ingredient that makes them natural liars.

If we extrapolate these findings to the general population, we might say that higher brain connectivity could make it easier for any of us to lie and at the same time think of ourselves as honorable creatures. After all, more connected brains have more avenues to explore when it comes to interpreting and explaining dubious events—and perhaps this is a crucial element in the rationalization of our dishonest acts.

More Creativity Equals More Money

These findings made me wonder whether increased white matter could be linked to both increased lying and increased creativity. After all, people who have more connections among their different brain parts and more associations are presumably also more creative. To test this possible link between creativity and dishonesty, Francesca Gino and I carried out a series of studies. True to the nature of creativity itself, we approached the question from a variety of angles, starting with a relatively simple approach.

When our participants showed up at the lab, we informed them that they would answer some questions followed by a computerized task. The question set included many irrelevant questions about their general experiences and habits (these filler questions were designed to obscure the real intent of the study) and three types of questions that were the focus of the study.

In the first set of questions, we asked the participants to indicate to what degree they would describe themselves using some "creative" adjectives (insightful, inventive, original, resourceful, unconventional, and so on). In the second, we asked them to tell us how often they engage in seventy-seven different activities, some of which require more creativity and some less (bowling, skiing, skydiving, painting, writing, and so forth). In the third and last set of questions, we asked participants to rate how much they identified with statements such as "I have a lot of creative ideas," "I prefer tasks that enable me to think creatively," "I like to do things in an original way," and other similar statements.

Once the participants completed the personality measures, we asked them to complete the dots task, which was presumably unconnected to the questions. In case you don't recall this task, flip back to pages 127–28 in chapter 5, "Why Wearing Fakes Make Us Cheat More."

What do you think happened? Would participants who chose a large number of creative adjectives, engaged in creative activities more frequently, and viewed themselves as more creative cheat more, less, or about the same as the participants who were not as creative?

We found that participants who clicked the more-on-right button (the one with the higher payout) more often tended to be the same people who scored higher on all three creativity measures. Moreover, the difference between more and less creative individuals was most pronounced in the cases where the difference in the number of dots on the right and left sides was relatively small.

This suggested that the difference between creative and less creative individuals comes into play mostly when there is ambiguity in the situation at hand and, with it, more room for justification. When there was an obvious difference between the number of dots on the two sides of the diagonal, the participants simply had to decide whether to lie or not. But when the trials were more ambiguous and it was harder to tell if there were more dots to the right or the left of the diagonal, creativity kicked into action—along with more cheating. The more creative the individuals, the better they were at explaining to themselves why there were more dots to the right of the diagonal (the side with the higher reward).

Put simply, the link between creativity and dishonesty seems related to the ability to tell ourselves stories about how we are doing the right thing, even when we are not. The more creative we are, the more we are able to come up with good stories that help us justify our selfish interests.

Does Intelligence Matter?

Although this was an intriguing result, we didn't get too excited just yet. This first study showed that creativity and dishonesty are correlated, but that doesn't necessarily mean that creativity is directly linked to dishonesty. For example, what if a third factor such as intelligence was the factor linked to both creativity and dishonesty?

The link among intelligence, creativity, and dishonesty seems especially plausible when one considers how clever people such as the Ponzi schemer Bernie Madoff or the famous check forger Frank Abagnale (the author of *Catch Me If You Can*) must have been to fool so many people. And so our next step was to carry out an experiment in which we checked to see whether creativity or intelligence was a better predictor of dishonesty.

Again, picture yourself as one of our participants. This time, the testing starts before you even set foot in the lab. A week earlier, you sit down at your home computer and complete an online survey, which includes questions to assess your creativity and also measure your intelligence. We measure your creativity using the same three measures from the previous study, and measure your intelligence in two ways. First, we ask you to answer three questions designed to test your reliance on logic versus intuition using a set of three questions collected by Shane Frederick (a professor at Yale University). Along with the correct answer, each question comes with an intuitive answer that is in fact incorrect.

To give you an example, try this one: "A bat and a ball cost $1.10 in total. The bat costs $1.00 more than the ball. How much does the ball cost?"

Quick! What's the answer?

Ten cents?

Good try, but no. It's the seductive answer, but not the right one.

Although your intuition prods you to answer "$0.10," if you rely on logic more than intuition, you'll check your answer just to be sure: "If the ball were \$0.10, the bat would be \$1.10, which combine to equal \$1.20, not \$1.10 (.1 + (1 + .1) = 1.2)! Once you realize that your initial instinct is wrong, you enlist your memory of high school algebra and produce the correct solution (.05 + (1 + .05) = 1.1): 5 cents. Doesn't it feel like the SATs all over again? And congratulations if you got it right. (If not, don't worry, you would have most likely aced the two other questions on this short test.)

Figure 4

The Cognitive Reflection Test (CRT)

A bat and a ball cost \$1.10. The bat costs \$1.00 more than the ball. How much does the ball cost?

________ cents

If it takes 5 machines to make 5 widgets in 5 minutes, how long would it take 100 machines to make 100 widgets?

________ minutes

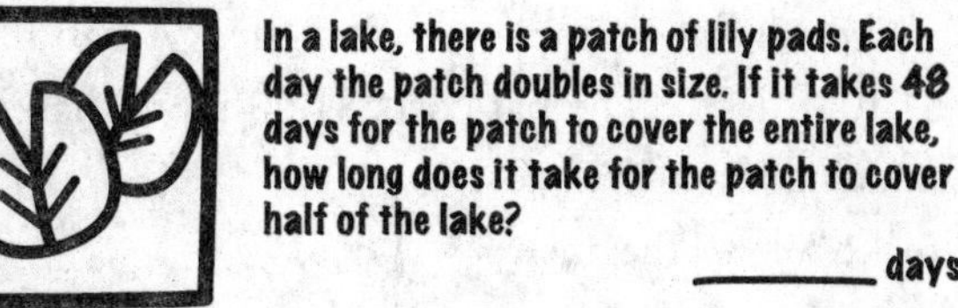

In a lake, there is a patch of lily pads. Each day the patch doubles in size. If it takes 48 days for the patch to cover the entire lake, how long does it take for the patch to cover half of the lake?

________ days

1) Default answer is 10; correct answer is 5.
2) Default answer is 100; correct answer is 5.
3) Default answer is 24; correct answer is 47.

Did you write your answers before, during, or after checking the answer key?

Next, we measure your intelligence through a verbal test. Here you're presented with a series of ten words (such as "dwindle" and "palliate"), and for each word you have to choose which of six options is closest in meaning to the target word.

A week later, you come to the lab and settle into one of the computer-facing chairs. Once you're situated, the instructions begin: "You'll be taking part in three different tasks today; these will test your problem-solving abilities, perceptual skills, and general knowledge. For the sake of convenience, we've combined them all into one session."

First up is the problem-solving task, which is none other than our trusty matrix task. When the five minutes for the test are over, you fold your worksheet and drop it into the recycling bin. What do you claim is your score? Do you report your actual score? Or do you dress it up a little?

Your second task, the perceptual skills task, is the dots test. Once again, you can cheat all you want. The incentive is there—you can earn $10 if you cheat in every one of the trials.

Finally, your third and final task is a multiple-choice general-knowledge quiz comprised of fifty questions of varying difficulty and subject matter. The questions include a variety of trivia such as "How far can a kangaroo jump?" (25 to 40 feet) and "What is the capital of Italy?" (Rome). For each correct answer, you receive 10 cents, for a maximum payout of $5. In the instructions for this last test, we ask that you circle your answers on the question sheet before later transferring them to a bubble sheet.

When you reach the end of this quiz, you put down your pencil. Suddenly the experimenter pipes up, "Oh, my gosh! I goofed! I mistakenly photocopied bubble sheets that are already marked with the correct answers. I'm so sorry. Would you mind using one of these premarked bubble sheets? I'll try to erase all the marks so that they will not show very clearly. Okay?" Of course you agree.

Next the experimenter asks you to copy your answers from the quiz to the premarked bubble sheet, shred the test sheets with your original answers, and only then submit the premarked bubble sheet with your answers in order to collect your payment. Obviously, as you transfer your answers you realize that you can cheat: instead of transferring your own answers to the bubble sheets, you can just fill in the premarked answers and take more money. ("I knew all along that the capital of Switzerland is Bern. I just chose Zurich without thinking about it.")

To sum things up, you've taken part in three tasks in which you can earn up to $20 to put toward your next meal, beer, or textbook. But how much you actually walk away with depends on your smarts and test-taking chops, as well as your moral compass. Would you cheat? And if so, do you think your cheating has anything to do with how smart you are? Does it have anything to do with how creative you are?

Here's what we found: as in the first experiment, the individuals who were more creative also had higher levels of dishonesty. Intelligence, however, wasn't correlated to any degree with dishonesty. This means that those who cheated more on each of the three tasks (matrices, dots, and general knowledge) had on average higher creativity scores compared to noncheaters, but their intelligence scores were not very different.

We also studied the scores of the extreme cheaters, the

participants who cheated almost to the max. In each of our measures of creativity, they had higher scores than those who cheated to a lower degree. Once again, their intelligence scores were no different.

Stretching the Fudge Factor: The Case for Revenge

Creativity is clearly an important means by which we enable our own cheating, but it's certainly not the only one. In an earlier book (*The Upside of Irrationality*) I described an experiment designed to measure what happens when people are upset by bad service. Briefly, Ayelet Gneezy (a professor at the University of California, San Diego) and I hired a young actor named Daniel to run some experiments for us in local coffee shops. Daniel asked coffee shop patrons to participate in a five-minute task in return for $5. When they agreed, he handed them ten sheets of paper covered with random letters and asked them to find as many identical adjacent letters as they could and circle them with a pencil. After they finished, he returned to their table, collected their sheets, handed them a small stack of bills, and told them, "Here is your $5, please count the money, sign the receipt, and leave it on the table. I'll be back later to collect it." Then he left to look for another participant. The key was that he gave them $9 rather than $5, and the question was how many of the participants would return the extra cash.

This was the no-annoyance condition. Another set of customers—those in the annoyance condition—experienced a slightly different Daniel. In the midst of explaining the task, he pretended that his cell phone was vibrating. He reached into his pocket, took out the phone, and said, "Hi,

Mike. What's up?" After a short pause, he would enthusiastically say, "Perfect, pizza tonight at eight thirty. My place or yours?" Then he would end his call with "Later." The whole fake conversation took about twelve seconds.

After Daniel slipped the cell phone back into his pocket, he made no reference to the disruption and simply continued describing the task. From that point on, everything was the same as in the no-annoyance condition.

We wanted to see if the customers who had been so rudely ignored would keep the extra money as an act of revenge against Daniel. Turns out they did. In the no-annoyance condition 45 percent of people returned the extra money, but only 14 percent of those who were annoyed did so. Although we found it pretty sad that more than half the people in the no-annoyance condition cheated, it was pretty disturbing to find that the twelve-second interruption provoked people in the annoyance condition to cheat much, much more.

In terms of dishonesty, I think that these results suggest that once something or someone irritates us, it becomes easier for us to justify our immoral behavior. Our dishonesty becomes retribution, a compensatory act against whatever got our goat in the first place. We tell ourselves that we're not doing anything wrong, we are only getting even. We might even take this rationalization a step further and tell ourselves that we are simply restoring karma and balance to the world. Good for us, we're crusading for justice!

MY FRIEND AND *New York Times* technology columnist David Pogue captured some of the annoyance we feel toward customer service—and the desire for revenge that comes with it. Anyone who knows David would tell you that he is the kind of person who would gladly help anyone in need, so the idea that he would go out of his way to hurt anyone is rather surprising—but when we feel hurt, there is hardly a limit to the extent to which we can reframe our moral code. And David, as you'll see in a moment, is a highly creative individual. Here is David's song (please sing along to the melody of "The Sounds of Silence"):

Hello voice mail, my old friend
I've called for tech support again
I ignored my boss's warning
I called on a Monday morning
Now it's evening and my dinner
First grew cold and then grew mold . . .
I'm still on hold!
I'm listening to the sounds of silence.

You don't seem to understand.
I think your phone lines are unmanned.
I punched every touchtone I was told,
But I've still spent 18 hours on hold.
It's not enough your program crashed my Mac
And it constantly hangs and bombs;
It erased my ROMs!
Now my Mac makes the sounds of silence.

In my dreams I fantasize
Of wreaking vengeance on you guys.
Say your motorcycle crashes;
Blood comes gushing from your gashes.
With your fading strength you call 911

And you pray for a trained MD . . .
But you get me!
And you listen to the sounds of silence!

An Italian Story of Creative Revenge

When I was seventeen and my cousin Yoav was eighteen, we spent the summer backpacking in Europe, having a wonderful time. We met lots of people, saw beautiful cities and places, spent time in museums—it was a perfect European jaunt for two restless teenagers.

Our travel itinerary went from Rome up through Italy and France and finally to England. When we originally bought our youth train passes, the nice fellow at the Rome Eurail office gave us a photocopy of a map of the European train system, carefully marking the train path that we were going to take with a black ballpoint pen. He told us that we could use our passes anytime we wanted within the two-month window but that we could travel only along the particular route he had drawn. He stapled the flimsy photocopied map to a more official printed receipt and handed us the package. Initially, we were certain that no conductor would respect this rather unsophisticated-looking map and ticket combo, but the ticket seller assured us that it was all we needed, and in fact that proved to be the case.

After enjoying the sights in Rome, Florence, Venice, and a few smaller Italian towns, we spent a few nights on the shore of a lake outside Verona. On our last night by the lake, we woke up to find that someone had been through our backpacks and strewn our stuff all over the place. After taking a careful inventory of our belongings, we saw that all of our clothes and even my camera were still there. The only thing missing was Yoav's extra pair of sneakers. We would have considered it a minor loss, except for the fact that Yoav's mother (my aunt Nava), in her infinite wisdom, had wanted to make sure that we had some emergency cash in case someone stole our money. So she had tucked a few hundred dollars in Yoav's extra pair of sneakers. The irony of the situation was painful.

We decided to look around the town to see if we could spot someone wearing Yoav's sneakers and went to the police as well. Given the fact that the local policemen understood little English, it was rather difficult to convey the nature of the crime—that a pair of sneakers had been stolen and that it was important because there was cash hidden in the sole of the right shoe. Not surprisingly, we never recovered Yoav's sneakers, and that left us somewhat embittered. In our minds it was an unfair turn of events, and Europe owed us one.

ABOUT A WEEK after the sneaker-theft incident, we decided that in addition to the other places on our route we also wanted to visit Switzerland and the Netherlands. We could have purchased new train tickets for the detour, but remembering the stolen shoes and the lack of help from the Italian police, we decided instead to expand our options with a bit of creativity. Using a black ballpoint pen just like the ticket seller's, we drew another path on our photocopied map. This one passed through Switzerland on the way to France and from there to England. Now the map showed two possible routes for our journey: the original route and our modified one. When we showed the maps to the next few conductors, they did not comment on our artwork, so we continued sketching extra routes on our maps for a few weeks.

Our scam worked until we were en route to Basel. The Swiss conductor examined our passes, scowled, shook his head, and handed them back to us.

"You are going to have to buy a ticket for this part of your trip," he informed us.

"Oh, but you see, sir," we said ever so politely, "Basel is indeed on our route." We pointed to the modified path on our map.

The conductor was unconvinced. "I am sorry, but you will have to pay for your ticket to Basel, or I will have to ask you to leave the train."

"But, sir," we argued, "all the other conductors have accepted our tickets with no problem."

The conductor shrugged and shook his head again.

"Please, sir," pleaded Yoav, "if you allow us to get to Basel, we will give you this tape of the Doors. They're a great American rock band."

The conductor did not seem amused or particularly interested in the Doors. "Okay," he said. "You can go to Basel."

We weren't sure whether he finally agreed with us, appreciated the gesture, or had just given up. After that incident we stopped adding lines to our map, and soon we returned to our original planned path.

LOOKING BACK ON our dishonest behavior, I am tempted to chalk it up to the stupidity of youth. But I know that's not the whole picture. In fact, I suspect that there are a number of aspects of the situation that enabled us to behave that way and justify our actions as perfectly acceptable.

First, I'm sure that being in a foreign country by ourselves for the first time helped us feel more comfortable with the new rules we were creating.* If we had stopped to give our actions more thought, we would have certainly recognized their seriousness, but somehow without thinking much, we imagined that our creative route enhancements were part of the regular Eurail procedure. Second, losing a few hundred dollars and Yoav's sneakers made us feel that it was okay for us to take some revenge and get Europe to pay us back. Third, since we were on an adventure, maybe we felt more morally adventurous too. Fourth, we justified our actions by convincing ourselves that we weren't really hurting anything or anyone. After all, we were just drawing a few extra lines on a piece of paper. The train was going on its track anyway; and besides, the trains were never full, so we weren't displacing anyone. It was also the case that we very easily justified our actions to ourselves because when we originally purchased the tickets, we could have picked a different route for the same price. And since the different paths were the same to the Eurail office when we originally purchased the tickets, why would it matter at what point in time we decided to choose a different path? (Maybe that's how people who backdate stock options justify their actions to themselves.) A final source of justification had to do with the physical nature of the ticket itself. Because the Eurail ticket seller had given us

* I suspect that there is a connection between dishonesty and traveling in general. Perhaps it's because when traveling the rules are less clear, or maybe it has to do with being away from one's usual setting.

just a flimsy piece of photocopied paper with a hand drawing of our planned route, it was physically easy for us to make our changes—and because we were just marking the path in the same way as the ticket seller (making lines on a piece of paper), this physical ease quickly translated into a moral ease as well.

When I think about all of these justifications together, I realize how extensive and expansive our ability to justify is and how prevalent rationalization can be in just about every one of our daily activities. We have an incredible ability to distance ourselves in all kinds of ways from the knowledge that we are breaking the rules, especially when our actions are a few steps removed from causing direct harm to someone else.

The Cheater's Department

Pablo Picasso once said, "Good artists copy, great artists steal." Throughout history, there has been no dearth of creative borrowers. William Shakespeare found his plot ideas in classical Greek, Roman, Italian, and historical sources and then wrote brilliant plays based on them. Even Steve Jobs occasionally boasted that much like Picasso, Apple was shameless about stealing great ideas.

Our experiments thus far suggested that creativity is a guiding force when it comes to cheating. But we didn't know whether we could take some people, increase their creativity, and with it also increase their level of dishonesty. This is where the next step of our empirical investigation came in.

In the next version of our experiments, Francesca and I looked into whether we could increase the level of cheating simply by getting our participants into a more creative mind-set (using what social scientists call priming). Imagine that you're one of the participants. You show up, and we introduce you to the dots task. You start off by completing a practice round, for which you do not receive any payment. Before you transition into the real phase—the one that involves the biased payment—we ask you to complete a sentence creation task. (This is where we work our creativity-inducing magic by using a scrambled sentence task, a common tactic for changing participants' momentary mind-sets.) In this task, you are given twenty sets of five words presented in a random order (such as "sky," "is," "the," "why," "blue"), and you are asked to construct a grammatically correct four-word sentence from each set ("The sky is blue"). What you don't know is that there are two different versions of this task, and you are going to see only one of them. One version is the creative set, in which twelve of the twenty sentences include words related to creativity ("creative," "original," "novel," "new," "ingenious," "imagination," "ideas," and so on). The other version is the control set, in which none of the twenty sentences includes any words related to creativity. Our aim was to prime some of the participants into a more innovative, aspiring mind-set à la Albert Einstein or Leonardo da Vinci by using the words associated with creativity. Everyone else was stuck with their usual mind-set.

Once you complete the sentence task (in one of the two versions), you go back to the dots task. But this time you're doing it for real money. Just as before, you earn half a cent for choosing the left side and 5 cents for choosing the right.

What kind of picture did the data paint? Did facilitating a more creative mind-set affect a person's morality? Although the two groups didn't differ in their levels of performance in the practice rounds of the dots task (when there was no payment), there was a difference after the scrambled sentence task. As we expected, the participants who had been primed with the creative words chose "right" (the response with the higher pay) more often than those in the control condition.

SO FAR, IT appeared that a creative mind-set could make people cheat a bit more. In the final stage of our investigation, we wanted to see how creativity and cheating correlate in the real world. We approached a large advertising agency and got most of the employees to answer a series of questions about moral dilemmas. We asked questions such as "How likely would you be to inflate your business expense report?"; "How likely would you be to tell your supervisor that progress has been made on a project when none has been made at all?"; and "How likely are you to take home office supplies from work?" We also asked them which department they worked for within the company (accounting, copywriting, account management, design, and so on). Finally, we got the CEO of the advertising agency to tell us how much creativity was required to work in each of the departments.

Now we knew the basic moral disposition of each employee, their departments, and the level of creativity expected in each department. With this data at hand, we computed the moral flexibility of the employees in each of the different departments and how this flexibility related to the creativity demanded by their jobs. As it turned out, the level of moral flexibility was highly related to the level of creativity required in their department and by their job. Designers and copywriters were at the top of the moral flexibility scale, and the accountants ranked at the bottom. It seems that when "creativity" is in our job description, we are more likely to say "Go for it" when it comes to dishonest behavior.

The Dark Side of Creativity

Of course, we're used to hearing creativity extolled as a personal virtue and as an important engine for the progress of society. It's a trait we aspire to—not just as individuals but also as companies and communities. We honor innovators, praise and envy those who have original minds, and shake our heads when others aren't able to think outside the box.

There's good reason for all of this. Creativity enhances our ability to solve problems by opening doors to new approaches and solutions. It's what has enabled mankind to redesign our world in (sometimes) beneficial ways with inventions ranging from sewer and clean water systems to solar panels, and from skyscrapers to nanotechnology. Though we still have a way to go, we can thank creativity for much of our progress. After all, the world would be a much bleaker place without creative trailblazers such as Einstein, Shakespeare, and da Vinci.

But that's only part of the story. Just as creativity enables us to envision novel solutions to tough problems, it can also enable us to develop original paths around rules, all the while allowing us to reinterpret information in a self-serving way. Putting our creative minds to work can help us come up with a narrative that lets us have our cake and eat it too, and create stories in which we're always the hero, never the villain. If the key to our dishonesty is our ability to think of ourselves as honest and moral people while at the same time benefitting from cheating, creativity can help us tell better stories—stories that allow us to be even more dishonest but still think of ourselves as wonderfully honest people.

The combination of positive and desired outcomes, on the one hand, and the dark side of creativity, on the other, leaves us in a tight spot. Though we need and want creativity, it is also clear that under some circumstances creativity can have a negative influence. As the historian (and also my colleague and friend) Ed Balleisen describes in his forthcoming book *Suckers, Swindlers, and an Ambivalent State*, every time business breaks through new technological frontiers—whether the invention of the postal service, the telephone, the radio, the computer, or mortgage-backed securities—such progress allows people to approach the boundaries of both technology and dishonesty. Only later, once the capabilities, effects, and limitations of a technology have been established, can we determine both the desirable and abusive ways to use these new tools.

For example, Ed shows that one of the first uses of the U.S. postal service was for selling products that did not exist. It took some time to figure that out, and eventually the problem of mail fraud ushered in a strong set of regulations that now help ensure the high quality, efficiency, and trust in this important service. If you think about technological development from this perspective, it means that we should be thankful to some of the creative swindlers for some of their innovation and some of our progress.

Where does this leave us? Obviously, we should keep hiring creative people, we should still aspire to be creative ourselves, and we should continue to encourage creativity in others. But we also need to understand the links between creativity and dishonesty and try to restrict the cases in which creative people might be tempted to use their skills to find new ways to misbehave.

BY THE WAY—I am not sure if I mentioned it, but I think that I am both incredibly honest and highly creative.

CHAPTER 8

Cheating as an Infection

How We Catch the Dishonesty Germ

I spend a lot of my time giving talks around the world about the effects of irrational behavior. So naturally, I'm a very frequent flyer. One typical itinerary included flying from my home in North Carolina to New York City, then on to São Paulo, Brazil; Bogotá, Colombia; Zagreb, Croatia; San Diego, California; and back to North Carolina. A few days later I flew to Austin, Texas; New York City; Istanbul, Turkey; Camden, Maine; and finally (exhausted) back home. In the process of accumulating all those miles, I've sustained an endless number of insults and injuries while grinding through security checkpoints and attempting to retrieve lost baggage. But those pains have been nothing compared to the pain of getting sick while traveling, and I am always trying to minimize my chances of falling ill.

On one particular transatlantic flight, while I was preparing a talk to give the next day on conflicts of interest, my neighbor seemed to have a bad cold. Maybe it was his sickness, my fear of catching something in general, sleep deprivation, or just the random and amusing nature of free associations that made me wonder about the similarity between the germs my seatmate and I were passing back and forth and the recent spread of corporate dishonesty.

As I've mentioned, the collapse of Enron spiked my interest in the phenomenon of corporate cheating—and my interest continued to grow following the wave of scandals at Kmart, WorldCom, Tyco, Halliburton, Bristol-Myers Squibb, Freddie Mac, Fannie Mae, the financial crisis of 2008, and, of course, Bernard L. Madoff Investment Securities. From the sidelines, it seemed that the frequency of financial scandals was increasing. Was this due to improvements in the detection of dishonest and illegal behavior? Was it due to a deteriorating moral compass and an actual increase in dishonesty? Or was there also an infectious element to dishonesty that was getting a stronger hold on the corporate world?

Meanwhile, as my sniffling neighbor's pile of used tissues grew, I began wondering whether someone could become infected with an "immorality bug." If there was a real increase in societal dishonesty, could it be spreading like an infection, virus, or communicable bacteria, transmitted through mere observation or direct contact? Might there be a connection between this notion of infection and the continually unfolding story of deception and dishonesty that we have increasingly seen all around us? And if there were such a connection, would it be possible to detect such a "virus" early and prevent it from wreaking havoc?

To me, this was an intriguing possibility. Once I got home, I started reading up on bacteria, and I learned that we have innumerable bacteria in, on, and around our bodies. I also learned that as long as we have only a limited amount of the harmful bacteria, we manage rather well. But problems tend to arise when the number of bacteria becomes so great that it disturbs our natural balance or when a particularly bad strain of bacteria makes it through our bodies' defenses.

To be fair, I am hardly the first to think of this connection. In the eighteenth and nineteenth centuries, prison reformers believed that criminals, like the ill, should be kept separated and in well-ventilated places in order to avoid contagion. Of course, I didn't take the analogy between the spread of dishonesty and diseases as literally as my predecessors. Some sort of airborne miasma probably won't transform people into criminals. But at the risk of overstretching the metaphor, I thought that the natural balance of social honesty could be upset, too, if we are put into close proximity to someone who is cheating. Perhaps observing dishonesty in people who are close to us might be more "infectious" than observing the same level of dishonesty in people who aren't so close or influential in our lives. (Consider, for example, the catchphrase "I learned it by watching you" from the antidrug campaign of the 1980s: the ad warned that "Parents who use drugs have children who use drugs.")

Keeping with the infection metaphor, I wondered about the intensity of exposure to cheating and how much dishonest behavior it might take to tilt the scale of our own actions. If we see a colleague walking out of the office supply room with a handful of pens, for example, do we immediately start thinking that it's all right to follow in his footsteps and grab some office supplies ourselves? I suspect that this is not the case. Instead, much like our relationship with bacteria, there might be a slower and more subtle process of accretion: perhaps when we see someone cheat, a microscopic impression is left with us and we become ever so slightly more corrupt. Then, the next time we witness unethical behavior, our own morality erodes further, and we become more and more compromised as the number of immoral "germs" to which we are exposed increases.

A FEW YEARS ago I purchased a vending machine, thinking it would be an interesting tool for running experiments related to pricing and discounts. For a few weeks, Nina Mazar and I used it to see what would happen if we gave people a probabilistic discount instead of a fixed discount. Translated, that means that we set up the machine so that some candy slots were marked with a 30 percent discount off the regular price of $1, while other slots gave users a 70 percent chance of paying the full price of $1.00 and a 30 percent chance of getting all their money back (and therefore paying nothing). In case you are interested in the results of this experiment, we almost tripled sales by probabilistically giving people back their money. This probabilistic discounting is a story for another time, but the idea of people getting their money back gave us an idea for testing another path for cheating.

One morning, I had the machine moved near a classroom building at MIT and set the internal price of the machine to zero for each of the candies. On the outside, each candy allegedly cost 75 cents. But the moment students shelled out three quarters and made their selection, the machine served both the candy and the money. We also put a prominent sign on the machine with a number to call if the machine malfunctioned.

A research assistant sat within eyeshot of the machine and pretended to work on her laptop. But instead she recorded what people did when confronted with the surprise of free candy. After doing this for a while, she observed two types of behavior. First, people took approximately three pieces of candy. When they got their first candy together with their payment, most people checked to see whether it would happen again (which, of course, it did). And then many people decided to go for it a third time. But no one tried more often than that. People undoubtedly remembered a time when a vending machine ate their money without dispensing anything, so they probably felt as though this generous machine was evening out their vending-machine karma.

We also found that more than half of the people looked around for a friend, and when they saw someone they knew, they invited their friend over to partake in the sugar-laden boon. Of course, this was just an observational study, but it led me to suspect that when we do something questionable, the act of inviting our friends to join in can help us justify our own questionable behavior. After all, if our friends cross the ethical line with us, won't that make our action seem more socially acceptable in our own eyes? Going to such lengths to justify our bad behavior might seem over the top, but we often take comfort when our actions fall in line with the social norms of those around us.

Infectious Cheating in Class

After my experience with the vending machine, I started observing the infectious nature of cheating in other places as well—including in my own classes. At the start of the semester a few years ago, I asked the five hundred undergraduate students in my behavioral economics class how many of them believed that they could listen carefully in class while using their computers for non-class-related activities (Facebook, Internet, e-mail, and so on). Thankfully, most of them indicated that they couldn't really multitask very well (which is true). I then asked how many of them had enough self-control to avoid using their laptop for non-class-related activities if it was open in front of them. Almost no one raised a hand.

At that point I was conflicted between prohibiting laptops in the classroom (which are of course useful for taking notes) or allowing laptops but, to help the students fight their lack of self-control, adding some intervention. Being an optimist, I asked the students to raise their right hands and repeat after me, "I will never, never, never use my computer in this course for anything that is not class-related. I will not read or send e-mail; I will not use Facebook or other social networks; and I will not use the Internet to explore any non-class-related material during class time."

The students repeated these words after me, and I was rather pleased with myself—for a while.

From time to time I show videos in class both to illustrate a point and to give the students a change in pace and attention. I usually take this time to walk to the back of the class and watch the videos with the students from there. Of course, standing at the back of the class also gives me a direct line of sight to the screens of the students' laptops. During the first few weeks of the semester their screens shone only with class-related material. But as the semester drew on—like mushrooms after the rain—I noticed that every week more and more of the screens were opened to very familiar but non-class-related websites and that Facebook and e-mail programs were often front and center.

In retrospect, I think that the darkness that accompanied the videos was one of the culprits in the deterioration of the students' promise. Once the class was in darkness and one student used his laptop for a non-class-related activity, even for just one minute, many of the other students, not just me, could see what he was doing. That most likely led more students to follow the same pattern of misbehavior. As I discovered, the honesty pledge was helpful in the beginning, but ultimately it was no match for the power of the emerging social norm that came from observing the misbehavior of others.*

One Bad Apple

My observations of on-campus cheating and my 30,000-foot musings about social infection were, of course, just speculations. To acquire a more informed view of the infectious nature of cheating, Francesca Gino, Shahar Ayal (a professor at the Interdisciplinary Center in Israel), and I decided to set up a few experiments at Carnegie Mellon University, where Francesca was visiting at the time. We set up the matrix task in the same general way I described earlier (although we used an easier version of the task), but with a few important differences. The first was that along with the worksheets containing the matrices, the experimenter handed out a manila envelope containing $10 worth of cash (eight $1 bills and four half-dollar coins) to each participant. This change in payment procedure meant that at the end of the experiment,

---

* The smart thing would have been to lead the students through the oath at the start of every lecture, and maybe this is what I will do next time.

the participants paid themselves and left behind their unearned cash.

In the control condition, in which there was no opportunity for cheating, a student who solved seven questions in the allotted time counted how many problems she solved correctly, withdrew the appropriate amount of money from the manila envelope, and placed the money in her wallet. Then the participant handed the worksheet and envelope with the unearned cash back to the experimenter, who checked the worksheet, counted the remaining money in the envelope, and sent the student away with her earnings. So far, so good.

In the shredder condition, the instructions were a bit different. In this condition the experimenter told the participants, "After you count your answers, head over to the shredder at the back of the room, shred your questionnaire, then walk back to your seat and take the amount of money you have earned from the manila envelope. After that, you are free to leave. On your way out, toss the envelope with the unearned money into the box by the door." Then she told the participants to start on the test and began reading a thick book (to make it clear that no one was watching). After the five minutes were over, the experimenter announced that the time was up. The participants put down their pencils, counted the number of their correct answers, shredded their worksheets, walked back to their seat, paid themselves, and on their way out tossed their envelopes containing the leftover money into the box. Not surprisingly, we found that participants in the shredder condition claimed to have solved more matrices than those in the control condition.

These two conditions created the starting point from

which we could test what we really wanted to look at: the social component of cheating. Next, we took the shredder condition (in which cheating was possible) and added a social element to it. What would happen if our participants could observe someone else—a Madoff in the making—cheating egregiously? Would it alter their level of cheating?

Imagine that you are a participant in our so-called Madoff condition. You're seated at a desk, and the experimenter gives you and your fellow participants the instructions. "You may begin!" she announces. You dive into the problem set, trying to solve as many matrices as possible to maximize your earnings. About sixty seconds pass, and you're still on the first question. The clock is ticking.

"I've finished!" a tall, skinny, blond-haired guy says as he stands up and looks at the experimenter. "What should I do now?"

"Impossible," you think. "I haven't even solved the first matrix!" You and everyone else stare at him in disbelief. Obviously, he's cheated. Nobody could have completed all twenty matrices in less than sixty seconds.

"Go shred your worksheet," the instructor tells him. The guy walks to the back of the room, shreds his worksheet, and then says, "I solved everything, so my envelope for the extra money is empty. What should I do with it?"

"If you don't have money to return," the experimenter replies, unfazed, "put the empty envelope in the box, and you are free to go." The student thanks her, waves good-bye to everyone, and leaves the room smiling, having pocketed the entire amount. Having observed this episode, how do you react? Do you become outraged that the guy cheated and got away with it? Do you change your own moral behavior? Do you cheat less? More?

It may make you feel slightly better to know that the fellow who cheated so outrageously was an acting student named David, whom we hired to play this role. We wanted to see if observing David's outrageous behavior would cause the real participants to follow his example, catching the "immorality virus," so to speak, and start cheating more themselves.

Here's what we found. In the Madoff condition, our participants claimed to have solved an average of fifteen out of twenty matrices, an additional eight matrices beyond the control condition, and an additional three matrices beyond the shredder condition. In short, those in the Madoff condition paid themselves for roughly double the number of answers they actually got right.

Here's a quick summary:

Condition	Problems "solved" (out of 20)	Magnitude of cheating
Control (Cheating not possible)	7	0
Shredder (Cheating possible)	12	5
Madoff (Cheating possible)	15	8

THOSE RESULTS, THOUGH interesting, still don't tell us why the participants in the Madoff condition were cheating more. Given David's performance, participants could have made a quick calculation and said to themselves, "If he can cheat and get away with it, it must mean that I can do the same without any fear of getting caught." If this were the case, David's action would have changed participants' cost-benefit analysis by clearly demonstrating that in this experiment, they could cheat and get away with it. (This is the SMORC perspective that we described in chapter 1, "Testing the Simple Model of Rational Crime.")

A very different possibility is that David's actions somehow signaled to the other participants in the room that this type of behavior was socially acceptable, or at least possible, among their peers. In many areas of life, we look to others to learn what behaviors are appropriate and inappropriate. Dishonesty may very well be one of the cases where the social norms that define acceptable behavior are not very clear, and the behavior of others—David, in this case—can shape our ideas about what's right and wrong. From this perspective, the increased cheating we observed in the Madoff condition could be due not to a rational cost-benefit analysis, but rather to new information and mental revision of what is acceptable within the moral boundaries.

To examine which of the two possibilities better explains the increased cheating in the Madoff condition, we set up another experiment, with a different type of social-moral information. In the new setup, we wanted to see whether erasing any concern about being caught but without giving an enacted example of cheating would also cause participants to cheat more. We got David to work for us again, but this time he interjected a question as the experimenter was wrapping up the instructions. "Excuse me," he said to the experimenter in a loud voice, "Given these instructions, can't I just say I solved everything and walk away with all the cash? Is this okay?" After pausing for a few seconds, the experimenter answered, "You can do whatever you want." For obvious reasons, we called this the question condition. Upon hearing this exchange, participants quickly understood that in this experiment they could cheat and get away with it. If you were a participant, would this understanding encourage you to cheat more? Would you conduct a quick cost-benefit analysis and figure that you could walk away with some unearned dough? After all, you heard the experimenter say, "Do whatever you want," didn't you?

Now let's stop and consider how this version of the experiment can help us understand what happened in the Madoff condition. In the Madoff condition the participants were provided with a live example of cheating behavior, which provided them with two types of information: From a cost-benefit perspective, watching David walk out with all the money showed them that in this experiment there are no negative consequences to cheating. At the same time, David's action provided them with a social cue that people just like them seem to be cheating in this experiment. Because the Madoff condition included both elements, we couldn't tell if the increased cheating that followed was due to a reevaluation of the cost-benefit analysis, to the social cue, or to both.

This is where the question condition comes in handy. In this condition, only the first element (cost-benefit perspective) was present. When David asked the question and the experimenter confirmed that cheating was not only possible but also without a consequence, it became clear to the participants that cheating in this setup had no downside. And most important, the question condition changed the participants' understanding of the consequence without giving them a live example and social cue of someone from their social group who cheated. If the amount of cheating in the question condition were the same as in the Madoff condition, we would conclude that what caused the increased level of cheating in both conditions was most likely the information that there was no consequence to cheating. On the other hand, if the amount of cheating in the question condition were much lower than in the Madoff condition, we would conclude that what caused the extra-high level of cheating in the Madoff condition was the social signal—the realization that people from the same social group find it acceptable to cheat in this situation.

What do you think happened? In the question condition, our participants claimed to have solved an average of ten matrices—about three more matrices than in the control condition (which means they did cheat) but by about two fewer matrices than in the shredder condition and by five fewer than in the Madoff condition. After observing the experimenter telling David that he could do what he wanted, cheating actually *decreased*. That was the opposite of what would have happened if our participants had engaged solely in a rational cost-benefit analysis. Moreover, this result suggests that when we become aware of the possibility of immoral behavior, we reflect on our own morality (similar to the Ten Commandments and the honor code experiments in chapter 2, "Fun with the Fudge Factor"). And as a consequence, we behave more honestly.

A Fashion Statement

Although those results were promising, we still wanted to get more direct support and evidence for the idea that cheating might be socially contagious. So we decided to go into the fashion business. Well, sort of.

The structure of our next experiment was the same as in the Madoff condition: our actor stood up a few seconds into the experiment and announced that he had solved everything and so forth. But this time there was one fashion-related difference: the actor wore a University of Pittsburgh sweatshirt.

Let me explain. Pittsburgh has two world-class universities, the University of Pittsburgh (UPitt) and Carnegie Mellon University (CMU). Like many institutions of higher learning that are in close proximity, these two have a long-standing rivalry. This competitive spirit was just what we needed to further test our cheating-as-a-social-contagion hypothesis.

We conducted all of these experiments at Carnegie Mellon University, and all our participants were Carnegie Mellon students. In the basic Madoff condition, David had worn just a plain T-shirt and jeans and had therefore been assumed to be a Carnegie Mellon student, just like all the other participants. But in our new condition, which we named the "outsider-Madoff condition," David wore a blue-and-gold UPitt sweatshirt. This signaled to the other students that he was an outsider—a UPitt student—and not part of their social group; in fact, he belonged to a rival group.

The logic of this condition was similar to the logic of the question condition. We reasoned that if the increased cheating we observed in the Madoff condition was due to the realization that if David could cheat and get away with it, so could the other participants, and it would not matter if David was dressed as a CMU or a UPitt student. After all, the information that there were no negative consequences to egregious cheating was the same regardless of his outfit. On the other hand, if the increase in cheating in the Madoff condition was due to an emerging social norm that revealed to our participants that cheating was acceptable in their social group, this influence would operate only when our actor was part of their in-group (a Carnegie Mellon student) and not when he was a member of another, rival group (a UPitt student). The crucial element in this design, therefore, was the social link connecting David to the other participants: when he was dressed in a UPitt sweatshirt, would the CMU students continue to play copycat, or would they resist his influence?

To recap the results so far, here's what we saw: When cheating was possible in the shredder condition but not publicized by David, students claimed to have solved, on average, twelve matrices—five more than they did in the control condition. When David stood up wearing regular CMU attire in the Madoff condition, the participants claimed to have solved about fifteen matrices. When David asked a question about the possibility of cheating and he was assured that it was possible, participants claimed to have solved only ten matrices. And finally, in the outsider-Madoff condition (when David wore a UPitt sweatshirt), the students observing him cheat, claimed to have solved only nine matrices. They still cheated relative to the control condition (by about two matrices), but they cheated by about six fewer matrices than when David was assumed to be a part of their CMU social group.

Here's how our results looked:

Condition	Problems "solved" (out of 20)	Magnitude of cheating
Control (Cheating not possible)	7	0
Shredder (Cheating possible)	12	5
Madoff (Cheating possible)	15	8
Question (Cheating possible)	10	3
Outsider-Madoff (Cheating possible)	9	2

Together, these results show not only that cheating is common but that it is infectious and can be increased by observing the bad behavior of others around us. Specifically, it seems that the social forces around us work in two different ways: When the cheater is part of our social group, we identify with that person and, as a consequence, feel that cheating is more socially acceptable. But when the person cheating is an outsider, it is harder to justify our misbehavior, and we become more ethical out of a desire to distance ourselves from that immoral person and from that other (much less moral) out-group.

More generally, these results show how crucial other people are in defining acceptable boundaries for our own behavior, including cheating. As long as we see other members of our own social groups behaving in ways that are outside the acceptable range, it's likely that we too will recalibrate our internal moral compass and adopt their behavior as a model for our own. And if the member of our in-group happens to be an authority figure—a parent, boss, teacher, or someone else we respect—chances are even higher that we'll be dragged along.

In with the In-Crowd

It's one thing to get riled up about a bunch of college students cheating their university out of a few dollars (although even this cheating accumulates quickly); it's quite another when cheating is institutionalized on a larger scale. When a few insiders deviate from the norm, they infect those around them, who in turn infect those around them, and so on—which is what I suspect occurred at Enron in 2001, on Wall Street leading up to 2008, and in many other cases.

One can easily imagine the following scenario: A well-known banker named Bob at Giantbank engages in shady dealings—overpricing some financial products, delaying reporting losses until the next year, and so on—and in the process he makes boatloads of money. Other bankers at Giantbank hear about what Bob is up to. They go out to lunch and, over their martinis and steaks, discuss what Bob is doing. In the next booth, some folks from Hugebank overhear them. Word gets around.

In a relatively short time, it is clear to many other bankers that Bob isn't the only person to fudge some numbers. Moreover, they consider him as part of their in-group. To them, fudging the numbers now becomes accepted behavior, at least within the realm of "staying competitive" and "maximizing shareholder value."*

Similarly, consider this scenario: one bank uses its government bailout money to pay out dividends to its shareholders (or maybe the bank just keeps the cash instead of lending it). Soon, the CEOs of other banks start viewing this as appropriate behavior. It is an easy process, a slippery slope. And it's the kind of thing that happens all around us every day.

BANKING, OF COURSE, is not the only place this unfortunate kind of escalation takes place. You can find it anywhere, including governing bodies such as the U.S. Congress. One example of deteriorating social norms in the U.S. legislative halls involves political action committees (PACs). About thirty years ago, these groups were established as a way for members of Congress to raise money for their party and fellow lawmakers to use during difficult election battles. The money comes primarily from lobbyists, corporations, and special-interest groups, and the amounts they give are not restricted to the same degree as contributions to individual

* I suspect that companies that adapt the ideology of maximizing shareholder value above all else can use this motto to justify a broad range of misbehaviors, from financial to legal to environmental cheating. The fact that the compensation of the executives is linked to the stock price probably only increases their commitment to "shareholder value."

candidates. Aside from being taxed and having to be reported to the FEC, few restrictions are placed on the use of PAC money.

As you might imagine, members of Congress have gotten into the habit of using their PAC funds for a gamut of non-election-related activities—from babysitting bills to bar tabs, Colorado ski trips, and so on. What's more, less than half of the millions of dollars raised by PACs has gone to politicians actually running in elections; the rest is commonly put toward different perks: fund-raising, overhead, staff, and other expenses. As Steve Henn of the NPR show *Marketplace* put it, "PACs put the fun in fundraising."[1]

To deal with the misuse of PAC money, the very first law that Congress passed after the 2006 congressional election was intended to limit the discretionary spending of Congress members, forcing them to publicly disclose how they spent their PAC money. However, and somewhat predictably from our perspective, the legislation seemed to have no effect. Just a few weeks after passing the law, the congressmen were behaving as irresponsibly as they had before; some spent the PAC money going to strip clubs, blowing thousands of dollars on parties, and generally conducting themselves without a semblance of accountability.

How can this be? Very simple. Over time, as congressmen have witnessed fellow politicians using PAC funds in dubious ways, their collective social norm has taken a turn for the worse. Little by little, it's been established that PACs can be used for all kinds of personal and "professional" activities—and now the misuse of PAC funds is as common as suits and ties in the nation's capital. As Pete Sessions (a Republican congressman from Texas) responded when he was questioned about dropping several grand at the Forty Deuce in Las Vegas, "It's hard for me to know what is normal or regular anymore."[2]

You might suspect, given the polarization in Congress, that such negative social influences would be contained within parties. You might think that if a Democrat breaks the rules, his behavior would influence only other Democrats and that bad behavior by Republicans would influence only Republicans. But my own (limited) experience in Washington, D.C., suggests that away from the watchful eye of the media, the social practices of Democrats and Republicans (however disparate their ideologies) are much closer than we think. This creates the conditions under which the unethical behavior of any congressman can extend beyond party lines and influence other members, regardless of their affiliation.

ESSAY MILLS

In case you're unfamiliar with them, essay mills are companies whose sole purpose is to generate essays for high school and college students (in exchange for a fee, of course). Sure, they claim that the papers are intended to help the students write their own original papers, but with names such as eCheat.com, their actual purpose is pretty clear. (By the way, the tagline of eCheat.com was at one point "It's Not Cheating, It's Collaborating.")

Professors, in general, worry about essay mills and their impact on learning. But without any personal experience using essay mills and without any idea about what they really do or how good they are, it is hard to know how worried we should be. So Aline Grüneisen (the lab manager of my

research center at Duke University) and I decided to check out some of the most popular essay mills. We ordered a set of typical college term papers from a few of the companies, and the topic of the paper we chose was (surprise!) "Cheating."

Here is the task that we outsourced to the essay mills:

> *When and why do people cheat? Consider the social circumstances involved in dishonesty, and provide a thoughtful response to the topic of cheating. Address various forms of cheating (personal, at work, etc.) and how each of these can be rationalized by a social culture of cheating.*

We requested a twelve-page term paper for a university-level social psychology class, using fifteen references, formatted in American Psychological Association (APA) style, to be completed in two weeks. This was, to our minds, a pretty basic and conventional request. The essay mills charged us from $150 to $216 per paper in advance.

Two weeks later, what we received would best be described as gibberish. A few of the papers attempted to mimic APA style, but none achieved it without glaring errors. The citations were sloppy and the reference lists abominable—including outdated and unknown sources, many of which were online news stories, editorial posts, or blogs, and some that were simply broken links. In terms of the quality of the writing itself, the authors of all of the papers seemed to have a tenuous grasp of the English language and the structure of a basic essay. Paragraphs jumped clumsily from one topic to another and often collapsed into list form, counting off various forms of cheating or providing a long stream of examples that were never explained or connected to the thesis of the paper. Of the many literary affronts, we found the following gems:

Cheating by healers. Healing is different. There is harmless healing, when healers-cheaters and wizards offer omens, lapels, damage to withdraw, the husband-wife back and stuff. We read in the newspaper and just smile. But these days fewer people believe in wizards.

If the large allowance of study undertook on scholar betraying is any suggestion of academia and professors' powerful yearn to decrease scholar betraying, it appeared expected these mind-set would component into the creation of their school room guidelines.

By trusting blindfold only in stable love, loyalty, responsibility and honesty the partners assimilate with the credulous and naïve persons of the past.

The future generation must learn for historical mistakes and develop the sense of pride and responsibility for its actions.

At that point we were rather relieved, figuring that the day had not yet arrived when students can submit papers from essay mills and get good grades. Moreover, we concluded that if students did try to buy a paper from an essay mill, just like us, they would feel they had wasted their money and wouldn't try it again.

But the story does not end there. We submitted the essays we purchased to WriteCheck.com, a website that inspects papers for plagiarism, and found that half of the papers we received were largely copied from existing works. We decided to take action and contacted the essay mills to request our money back. Despite the solid proof from WriteCheck.com, the essay mills insisted that they had not plagiarized anything. One company even threatened us with litigation and claimed that they would get in touch with the dean's office at Duke to alert him to the fact that I had submitted work that was not mine. Needless to say, we never received that refund . . .

The bottom line? Professors shouldn't worry too much about essay mills, at least for now. The technological revolution has not yet solved this particular challenge for students, and they still have no other option but to write their own papers (or maybe cheat the old-fashioned way and use a paper from a student who took the class during a previous semester).

But I do worry about the existence of essay mills and the signal that they send to our students—that is, the institutional acceptance of cheating, not only while they are in school but after they graduate.

How to Regain Our Ethical Health?

The idea that dishonesty can be transmitted from person to person via social contagion suggests that we need to take a different approach to curbing dishonesty. In general, we tend to view minor infractions as just that: trivial and inconsequential. Peccadilloes may be relatively insignificant in and of themselves, but when they accumulate within a person, across many people, and in groups, they can send a signal that it's all right to misbehave on a larger scale. From this perspective, it's important to realize that the effects of individual transgressions can go beyond a singular dishonest act. Passed from person to person, dishonesty has a slow, creeping, socially erosive effect. As the "virus" mutates and spreads from one person to another, a new, less ethical code of conduct develops. And although it is subtle and gradual, the final outcome can be disastrous. This is the real cost of even minor instances of cheating and the reason we need to be more vigilant in our efforts to curb even small infractions.

So what can we do about it? One hint may lie in the Broken Windows Theory, which was the basis of a 1982 *Atlantic* article by George Kelling and James Wilson. Kelling and Wilson proposed a critical component of keeping order in dangerous neighborhoods, and it wasn't just putting more police on the beat. They argued that if people in a run-down area of town see a building with a few broken, long-unrepaired windows, they will be tempted to break even more windows and create further damage to the building and its surroundings, creating a blight effect. Based on the Broken Windows Theory, they suggested a simple strategy for preventing vandalism: fix problems when they are small. If you repair each broken window (or other misbehaviors) immediately, other potential offenders are going to be much less likely to misbehave.

Although the Broken Windows Theory has been difficult to prove or refute, its logic is compelling. It suggests that we should not excuse, overlook, or forgive small crimes, because doing so can make matters worse. This is especially important for those in the spotlight: politicians, public servants, celebrities, and CEOs. It might seem unfair to hold them to higher standards, but if we take seriously the idea that publicly observed behavior has a broader impact on those viewing the behavior, this means that their misbehavior can have greater downstream consequences for society at large. In contrast to this view, it seems that celebrities are too often rewarded with lighter punishments for their crimes than the rest of the population, which might suggest to the public that these crimes and misdemeanors are not all that bad.

THE GOOD NEWS is that we can also take advantage of the positive side of moral contagion by publicizing the individuals who stand up to corruption. For example, Sherron Watkins of Enron, Coleen Rowley of the FBI, and Cynthia Cooper of WorldCom are great examples of individuals who stood up to internal misconduct in their own organizations, and in 2002 *Time* magazine selected them as People of the Year.

Acts of honesty are incredibly important for our sense of social morality. And although they are unlikely to make the same sensational news, if we understand social contagion, we must also recognize the importance of publicly promoting outstanding moral acts. With more salient and vivid examples of commendable behavior, we might be able to improve what society views as acceptable and unacceptable behaviors, and ultimately improve our actions.

CHAPTER 9

Collaborative Cheating

Why Two Heads Aren't Necessarily Better than One

If you've ever worked in just about any organization, you know that working in teams accounts for a lot of your time. A great deal of economic activity and decision making takes place through collaboration. In fact, the majority of U.S. companies depend on group-based work, and more than half of all U.S. employees currently spend at least part of their day working in a group setting.[1] Try to count the number of meetings, project teams, and collaborative experiences you've had over the last six months, and you will quickly realize how many working hours these group activities consume. Group work also plays a prominent role in education. For example, the majority of MBA students' assignments consist of group-based tasks, and many undergraduate classes also require group-based projects.

In general, people tend to believe that working in groups has a positive influence on outcomes and that it increases the overall quality of decisions.[2] (In fact, much research has shown that collaboration can decrease the quality of decisions. But that's a topic for another time.) In general, the belief is that there is little to lose and everything to gain from collaboration—including encouraging a sense of camaraderie, increasing the level of fun at work, and benefitting from sharing and developing new ideas—all of which add up to more motivated and effective employees. What's not to love?

A FEW YEARS ago, in one of my graduate classes, I lectured about some of my research related to conflicts of interest (see chapter 3, "Blinded by Our Own Motivations"). After class, a student (I'll call her Jennifer) told me that the discussion had struck a chord with her. It reminded her of an incident that had taken place a few years earlier, when she was working as a certified public accountant (CPA) for a large accounting firm.

Jennifer told me that her job had been to produce the annual reports, proxy statements, and other documents that would inform shareholders about the state of their companies' affairs. One day her boss asked her to have her team prepare a report for the annual shareholders' meeting of one of their larger clients. The task involved going over all of the client's financial statements and determining the company's financial standing. It was a large responsibility, and Jennifer and her team worked hard to put together a comprehensive and detailed report that was honest and realistic. She did her best to prepare the report as accurately as possible, without, for example, overclaiming the company's profits or delaying reporting any losses to the next accounting year. She then left the draft of the report on her boss's desk, looking forward (somewhat anxiously) to his feedback.

Later that day, Jennifer got the report back with a note from her boss. It read, "I don't like these numbers. Please gather your team and get me a revised version by next Wednesday." Now, there are many reasons why her boss might not have "liked" the numbers, and it wasn't entirely clear to her what he meant. Moreover, not "liking" the numbers is an entirely different matter from the numbers being wrong—which was never implied. A multitude of questions ran through Jennifer's head: "What exactly did he want? How different should I make the numbers? Half a percent? One percent? Five percent?" She also didn't understand who was going to be accountable for any of the "improvements" she made. If the revisions turned out to be overly optimistic and someone was going to take the blame for it down the road, would it be her boss or her?

THE PROFESSION OF accounting is itself a somewhat equivocal trade. Sure, there are some clear-cut rules. But then there is a vaguely titled body of suggestions—known as Generally Accepted Accounting Principles (GAAP)—that accountants are supposed to follow. These guidelines afford accountants substantial leeway; they are so general that there's considerable variation in how accountants can interpret financial statements. (And often there are financial incentives to "bend" the guidelines to some degree.) For instance, one of the rules, "the principle of sincerity," states that the accountant's report should reflect the company's financial status "in good faith." That's all well and good, but "in good faith" is both excessively vague and extremely subjective. Of course, not everything (in life or accounting) is precisely quantifiable, but "in good faith" begs a few questions: Does it mean that accountants can act in bad faith?* And toward whom is this good faith directed? The people who run the company? Those who would like the books to look impressive and profitable (which would increase their bonuses and compensation)? Or should it be directed toward the people who have invested in the company? Or is it about those who want a clear idea of the company's financial condition?

Adding to the inherent complexity and ambiguity of her original task, Jennifer was now put under additional pressure by her boss. She'd prepared the initial report in what seemed to her to be good faith, but she realized that she was being asked to bend the accounting rules to some degree. Her boss wanted numbers that reflected more favorably upon the client company. After deliberating for a while, she concluded that she and her team should comply with his request; after all, he was her boss, and he certainly knew a lot more than she did about accounting, how to work with clients, and the client's expectations. In the end, although Jennifer started the process with every intention of being as accurate as possible, she wound up going back to the drawing board, reviewing the statements, reworking the numbers, and returning with a "better" report. This time, her boss was satisfied.

AFTER JENNIFER TOLD me her story, I continued to think about her work environment and the effect that working on a team with her boss and teammates had on her decision to

* Another fuzzy rule is the quaint-sounding "principle of prudence," according to which accountants should not make things appear rosier than they actually are.

push the accounting envelope a bit further. Jennifer was certainly in the kind of situation that people frequently face in the workplace, but what really stood out for me was that in this case the cheating took place in the context of a team, which was different from anything we had studied before.

In all of our earlier experiments on cheating, one person alone made the decision to cheat (even if he or she was spurred along by a dishonest act of another person). But in Jennifer's case, more than one person was directly involved, as is frequently the case in professional settings. In fact, it was clear to Jennifer that in addition to herself and her boss, her teammates would be affected by her actions. At the end of the year, the whole team would be evaluated together as a group—and their bonuses, raises, and future prospects were intertwined.

I started to wonder about the effects of collaboration on individual honesty. When we are part of a group, are we tempted to cheat more? Less? In other words, is a group setting conducive or destructive to honesty? This question is related to a topic we discussed in the previous chapter ("Cheating as an Infection"): whether it's possible that people can "catch" cheating from one another. But social contagion and social dependency are different. It's one thing to observe dishonest behavior in others and, based on that, alter our perceptions of what acceptable social norms are; it's quite another if the financial welfare of others depends on us.

Let's say you're working on a project with your coworkers. You don't necessarily observe them doing anything shady, but you know that they (and you) will benefit if you bend the rules a bit. Will you be more likely to do so if you know that they too will get something out of it? Jennifer's account suggests that collaboration can cause us to take a few extra liberties with moral guidelines, but is this the case in general?

Before we take a tour of some experiments examining the impact of collaboration on cheating, let's take a step back and think about possible positive and negative influences of teams and collaboration on our tendency to be dishonest.

Altruistic Cheating: Possible Costs of Collaboration

Work environments are socially complex, with multiple forces at play. Some of those forces might make it easy for group-based processes to turn collaborations into cheating opportunities in which individuals cheat to a higher degree because they realize that their actions can benefit people they like and care about.

Think about Jennifer again. Suppose she was a loyal person and liked to think of herself that way. Suppose further that she really liked her supervisor and team members and sincerely wanted to help them. Based on such considerations, she might have decided to fulfill her boss's request or even take her report a step further—not because of any selfish reasons but out of concern for her boss's well-being and deep regard for her team members. In her mind, "bad" numbers might get her boss and team members to fall out of favor with the client and the accounting company—meaning that Jennifer's concern for her team might lead her to increase the magnitude of her misbehavior.

Underlying this impulse is what social scientists call social utility. This term is used to describe the irrational but very human and wonderfully empathetic part of us that causes us to care about others and take action to help them out when we can—even at a cost to ourselves. Of course, we are all motivated to act in our own self-interest to some degree, but we also have a desire to act in ways that benefit those around us, particularly those we care about. Such altruistic feelings motivate us to help a stranger who is stuck with a flat tire, return a wallet we've found in the street, volunteer at a homeless shelter, help a friend in need, and so on.

This tendency to care about others can also make it possible to be more dishonest in situations where acting unethically will benefit others. From this perspective, we can think about cheating when others are involved as altruistic—where, like Robin Hood, we cheat because we are good people who care about the welfare of those around us.

Watch Out: Possible Benefits of Collaboration

In Plato's "Myth of the King of Gyges," a shepherd named Gyges finds a ring that makes him invisible. With this newfound power, he decides to go on a crime spree. So he travels to the king's court, seduces the queen, and conspires with her to kill the king and takes control of the kingdom. In telling the story, Plato wonders whether there is anyone alive who could resist taking advantage of the power of invisibility. The question, then, is whether the only force that keeps us from carrying out misdeeds is the fear of being seen by others (J. R. R. Tolkien elaborated on this theme a couple millennia later in *The Lord of the Rings*). To me, Plato's myth offers a nice illustration of the notion that group settings can inhibit our propensity to cheat. When we work within a team, other team members can act informally as monitors, and, knowing that we are being watched, we may be less inclined to act dishonorably.

A CLEVER EXPERIMENT by Melissa Bateson, Daniel Nettle, and Gilbert Roberts (all from the University of Newcastle) illustrated the idea that the mere feeling of being watched can inhibit bad behavior. This experiment took place in the kitchen of the psychology department at the University of Newcastle where tea, coffee, and milk were available for the professors and staff. Over the tea-making area hung a sign saying that beverage drinkers should contribute some cash to the honesty box located nearby. For ten weeks the sign was decorated with images, but the type of image alternated every week. On five of the weeks the sign was decorated with images of flowers, and on the other five weeks the sign was decorated with images of eyes that stared directly at the beverage drinkers. At the end of every week, the researchers counted the money in the honesty box. What did they find? There was some money in the box at the end of the weeks when the image of flowers was hung, but when the glaring eyes were "watching," the box contained almost three times more money.

As is the case with many findings in behavioral economics, this experiment produced a mix of good and bad news. On the negative side, it showed that even members of the psychology department—who you would think would know better—tried to sneak off without paying their share for a common good. On the positive side, it showed that the mere suggestion that they were being watched made them behave more honestly. It also shows that a full-blown Orwellian "Big Brother is watching" approach is not necessary and that much more subtle suggestions of being watched can be effective in increasing honesty. Who knows? Perhaps a warning sign, complete with watchful eyes, on Jennifer's boss's wall might have made a difference in his behavior.

IN PONDERING JENNIFER'S situation, Francesca Gino, Shahar Ayal, and I began to wonder how dishonesty operates in collaborative environments. Does monitoring help to reduce cheating? Do social connections in groups increase both altruism and dishonesty? And if both of these forces exert their influence in opposite directions, which of the two is more powerful? In order to shed light on this question, we turned once again to our favorite matrix experiment. We included the basic control condition (in which cheating was not possible), the shredder condition (in which cheating was possible), and we added a new condition that introduced a collaborative element to the shredder condition.

As our first step in exploring the effects of groups, we didn't want the collaborators to have an opportunity to discuss their strategy or to become friends, so we came up with a collaboration condition that included no familiarity or connection between the two team members. We called it the distant-group condition. Let's say you are one of the participants in the distant-group condition. As in the regular shredder condition, you sit at a desk and use a number 2 pencil to work on the matrices for five minutes. When the time is up, you walk to the shredder and destroy your test sheet.

Up to that point, the procedure is the same as in the basic shredder condition, but now we introduce the collaborative element. The experimenter tells you that you are part of a two-person team and that each of you will be paid half of the group's total earnings. The experimenter points out that your collection slip is either blue or green and has a number printed in the top-right corner. The experimenter asks you to walk around the room and find the person whose collection slip is different in color but with the same number in the top-right corner. When you find your partner, you sit down together, and each of you writes the number of matrices you correctly solved on your collection slip. Next, you write the other person's score on your collection slip. And finally, you combine the numbers for a total performance measure. Once that's done, you walk over to the experimenter together and hand him both collection slips. Since your worksheets have been shredded, the experimenter has no way to check the validity of your reported earnings. So he takes your word for it, pays you accordingly, and you split the takings.

Do you think people in this situation would cheat more than they did in the individual shredder condition? Here's what we found: when participants learned that both they and someone else would benefit from their dishonesty if they exaggerated their scores more, they ended up engaging in even higher levels of cheating, claiming to have solved three more matrices than when they were cheating just for themselves. This result suggests that we humans have a weakness for altruistic cheating, even if we barely know the person who might benefit from our misbehavior. Sadly, it seems that even altruism can have a dark side.

That's the bad news, and it's not all of it.

HAVING ESTABLISHED ONE negative aspect of collaboration—that people are more dishonest when others, even strangers, can benefit from their cheating—we wanted to turn our experimental sights on a possible positive aspect of collaboration and see what would happen when team members watch each other. Imagine that you're in a room with a few other participants, and you're randomly paired up with someone you have never met before. As luck would have it, you've ended up with a friendly-looking young woman. Before you have a chance to talk to her, you have to complete the matrix task in complete silence. You are player 1, so you start first. You tear into the first matrix, then the second, and then the third. All the while, your partner watches your attempts, successes, and failures. When the five minutes are up, you silently put your pencil down and your partner picks hers up. She starts working on her matrix task while you observe her progress. When the time is up, you walk to the shredder together and shred your worksheets. Then you each write down your own score on the same slip of paper, combine the two numbers for your joint performance score, and walk over to the experimenter's desk to collect your payment—all without saying a word to each other.

What level of cheating did we find? None at all. Despite the general inclination to cheat that we observe over and over, and despite the increase in the propensity to cheat when others can benefit from such actions, being closely supervised eliminated cheating altogether.

SO FAR, OUR experiments on cheating in groups showed two forces at play: altruistic tendencies get people to cheat more when their team members can benefit from their dishonesty, but direct supervision can reduce dishonesty and even eliminate it altogether. Given the coexistence of these two forces, the next question is: which force is more likely to overpower the other in more standard group interactions?

To answer this question, we needed to create an experimental setting that was more representative of how group members interact in a normal, day-to-day environment. You probably noticed that in the first two experiments, our participants didn't really interact with each other, whereas in daily life, group discussion and friendly chatter are an essential and inherent part of group-based collaborations. Hoping to add this important social element to our experimental setup, we devised our next experiment. This time, participants were encouraged to talk to each other, get to know each other, and become friendly. We even gave them lists of questions that they could ask each other in order to break the ice. They then took turns monitoring each other while each of them solved the matrices.

Sadly, we found that cheating reared its ugly head when we added this social element to the mix. When both elements were in the mix, the participants reported that they correctly solved about four extra matrices. So whereas altruism can increase cheating and direct supervision can decrease it, altruistic cheating overpowers the supervisory effect when people are put together in a setting where they have a chance to socialize and be observed.

LONG-TERM RELATIONSHIPS

Most of us tend to think that the longer we are in a relationship with our doctors, accountants, financial advisers, lawyers, and so on, the more likely it is that they will care more deeply about our well-being, and as a consequence, they will more likely put our needs ahead of their own. For

example, imagine that you just received a (nonterminal) diagnosis from your physician and you are faced with two treatment options. One is to start an aggressive, expensive therapy; the other is to wait awhile and see how your body deals with the problem and how it progresses ("watchful waiting" is the official term for this). There is not a definite answer as to which of the two options is better for you, but it is clear that the expensive, aggressive one is better for your physician's pocket. Now imagine that your physician tells you that you should pick the aggressive treatment option and that you should schedule it for next week at the latest. Would you trust his advice? Or would you take into account what you know about conflicts of interests, discount his advice, and maybe go for a second opinion? When faced with such dilemmas, most people trust their service providers to a very high degree and we are even more likely to trust them the longer we have known them. After all, if we have known our advisers for many years, wouldn't they start caring about us more? Wouldn't they see things from our perspective and give us better advice?

Another possibility, however, is that as the relationship extends and grows, our paid advisers—intentionally or not—become more comfortable recommending treatments that are in their own best interest. Janet Schwartz (the Tulane professor who, along with me, enjoyed dinner with the pharmaceutical reps), Mary Frances Luce (a professor at Duke University), and I tackled this question, sincerely hoping that as relationships between clients and service providers deepened, professionals would care more about their clients' welfare and less about their own. What we found, however, was the opposite.

We examined this question by analyzing data from millions of dental procedures over twelve years. We looked at instances when patients received fillings and whether the fillings were made of silver amalgam or white composite. You see, silver fillings last longer, cost less, and are more durable; white fillings, on the other hand, are more expensive and break more easily but are more aesthetically pleasing. So when it comes to our front teeth, aesthetics often reign over practicality, making white fillings the preferred option. But when it comes to our less visible back teeth, silver fillings are the way to go.[3]

What we found was that about a quarter of all patients receive attractive and expensive white fillings in their hidden teeth rather than the functionally superior silver fillings. In those cases, it was most likely that the dentists were making decisions that favored their own interests (higher initial pay and more frequent repairs) over the patients' interests (lower cost and longer-lasting treatment).

As if that weren't bad enough, we also found that this tendency is more pronounced the longer the patient sees the same dentist (we found the same pattern of results for other procedures as well). What this suggests is that as dentists become more comfortable with their patients, they also more frequently recommend procedures that are in their own financial interest. And long-term patients, for their part, are more likely to accept the dentist's advice based on the trust that their relationship has engendered.*

* Are dentists doing this on purpose, and do the patients know that they are being punished for their loyalty? Most likely it is not intentional, but whether conscious or not, the problem remains.

The bottom line: there are clearly many benefits to continuity of care and ongoing patient-provider relationships. Yet, at the same time, we should also be aware of the costs these long-term relationships can have.

HERE'S WHAT WE'VE learned about collaborative cheating so far:

Figure 5

Lessons in Collaborative Cheating

When we are working with a distant and unknown partner who might benefit from our cheating, we are more likely to cheat than if were cheating only for ourselves.

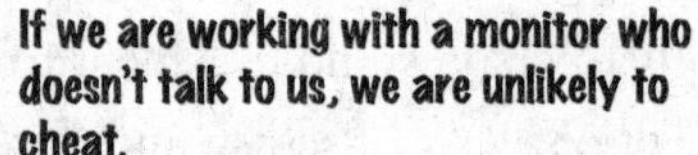

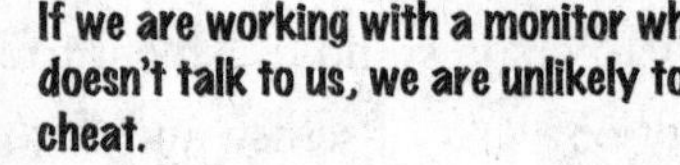

If we are working with a monitor who doesn't talk to us, we are unlikely to cheat.

If we are working with a monitor with whom we are becoming friendly, we are more likely to cheat than if we were cheating for a person we don't know that well.

In the end, it appears that the social aspects of cheating are so powerful that they can trump the beneficial effects of monitoring.

BUT WAIT, THERE'S MORE! In our initial experiments, both the cheater and the partner benefited from every additional exaggeration of their score. So if you were the cheater in the experiment and you exaggerated the number of your correct responses by one, you would get half of the additional payment and your partner would get the same. This is certainly less financially rewarding than snagging the whole amount for yourself, but you would still benefit from your exaggeration to some degree.

To look into purely altruistic cheating, we introduced a condition in which the fruit of each participant's cheating would benefit *only* their partner. What did we find? As it turns out, altruism is indeed a strong motivator for cheating. When cheating was carried out for purely altruistic reasons and the cheaters themselves did not gain anything from their act, overclaiming increased to an even larger degree.

Why might this be the case? I think that when both we and another person stand to benefit from our dishonesty, we operate out of a mix of selfish and altruistic motives. In contrast, when other people, and only other people, stand to benefit from our cheating, we find it far easier to rationalize our bad behavior in purely altruistic ways and subsequently we further relax our moral inhibitions. After all, if we are doing something for the pure benefit of others, aren't we indeed a little bit like Robin Hood?*

* Based on these results, we could speculate that people who work for ideological organizations such as political groups and not-for-profits might actually feel more comfortable bending moral rules—because they are doing it for a good cause and to help others.

FINALLY, IT IS worthwhile to say something more explicit about performance in the many control conditions that we had in this set of experiments. For each of our cheating conditions (individual shredder, group with shredder, distant group with shredder, friendly group with shredder, altruistic payoff with shredder), we also had a control condition in which there was no opportunity to cheat (that is, no shredder). Looking across these many different control conditions allowed us to see if the nature of collaboration influenced the level of performance. What we found was that performance was the same across all of these control conditions. Our conclusion? It seems that performance doesn't necessarily improve when people work in groups—at least not as much as we've all been led to believe.

OF COURSE, WE cannot survive without the help of others. Working together is a crucial element of our lives. But clearly, collaboration is a double-edged sword. On the one hand, it increases enjoyment, loyalty, and motivation. On the other hand, it carries with it the increased potential for cheating. In the end—and very sadly—it may be that the people who care the most about their coworkers end up cheating the most. Of course, I am not advocating that we stop working in groups, stop collaborating, or stop caring about one another. But we do need to recognize the potential costs of collaboration and increased affinity.

The Irony of Collaborative Work

If collaboration increases dishonesty, what can we do about it? One obvious answer is to increase monitoring. In fact, this seems to be the default response of the government's regulators to every instance of corporate misconduct. For example, the Enron fiasco brought about a large set of reporting regulations known as the Sarbanes-Oxley Act, and the financial crisis of 2008 ushered in an even larger set of regulations (largely emerging from the Dodd-Frank Wall Street Reform and Consumer Protection Act), which were designed to regulate and increase the supervision of the financial industry.

To some degree, there is no question that monitoring can be helpful, but it is also clear from our results that increased monitoring alone is unlikely to completely overcome our ability to justify our own dishonesty—particularly when others stand to gain from our misbehavior (not to mention the high financial costs of compliance with such regulations).

In some cases, instead of adding layers and layers of rules and regulations, perhaps we could set our sights on changing the nature of group-based collaboration. An interesting solution to this problem was recently implemented in a large international bank by a former student of mine named Gino. To allow his team of loan officers to work together without risking increased dishonesty (for example, by recording the value of the loans as higher than they really were in an effort to show larger short-run profits), he set up a unique supervisory system. He told his loan officers that an outside group would review their processing and approval of loan applications. The outside group was socially disconnected from the loan-making team and had no loyalty or motivation to help out the loan officers. To make sure that the two groups were separated, Gino located them in different office buildings. And he ensured that they had no direct dealings with each other or even knew the individuals in the other group.

I tried to get the data from Gino in order to evaluate the success of his approach, but the lawyers of this large bank stopped us. So, I don't know whether this approach worked or how his employees felt about the arrangement, but I suspect that this mechanism had at least some positive outcomes. It probably decreased the fun that the loan work group had during their meetings. It likely also increased the stress surrounding the groups' decisions, and it was certainly not cheap to implement. Nevertheless, Gino told me that overall, adding the objective and anonymous monitoring element seemed to have a positive effect on ethics, morals, and the bottom line.

CLEARLY, THERE ARE no silver bullets for the complex issue of cheating in group settings. Taken together, I think that our findings have serious implications for organizations, especially considering the predominance of collaborative work in our day-to-day professional lives. There is also no question that better understanding the extent and complexity of dishonesty in social settings is rather depressing. Still, by understanding the possible pitfalls involved in collaboration, we can take some steps toward rectifying dishonest behavior.

CHAPTER 10

A Semioptimistic Ending

People Don't Cheat Enough!

Throughout this book, we've seen that honesty and dishonesty are based on a mixture of two very different types of motivation. On the one hand, we want to benefit from cheating (this is the rational economic motivation), while on the other, we want to be able to view ourselves as wonderful human beings (this is the psychological motivation). You might think that we can't achieve both of these objectives at the same time—that we can't have our cake and eat it too, so to speak—but the fudge factor theory we have developed in these pages suggests that our capacity for flexible reasoning and rationalization allows us to do just that. Basically, as long as we cheat just a little bit, we can have the cake and eat (some of) it too. We can reap some of the benefits of dishonesty while maintaining a positive image of ourselves.

As we've seen, certain forces—such as the amount of money we stand to gain and the probability of being caught—influence human beings surprisingly less than one might think. And at the same time other forces influence us more than we might expect: moral reminders, distance from money, conflicts of interest, depletion, counterfeits, reminders of our fabricated achievements, creativity, witnessing others' dishonest acts, caring about others on our team, and so on.

ALTHOUGH THE FOCUS of the various experiments presented here was on dishonesty, it is also important to remember that most of the participants in our experiments were nice people from good universities who will likely attain positions of some power and influence later on in life. They were not the kind of people one typically associates with cheating. In fact, they were just like you, me, and most of the people on this planet, which means that all of us are perfectly capable of cheating a little bit.

Though that may sound pessimistic, the half-full part of the story is that human beings are, by and large, more moral than standard economic theory predicts. In fact, seen from a purely rational (SMORC) perspective, we humans don't cheat nearly enough. Consider how many times in the last few days you've had the opportunity to cheat without getting caught. Perhaps a colleague left her purse on her desk while she was away for a long meeting. Maybe a stranger in a coffee shop asked you to watch her laptop while she went to the restroom. Maybe a grocery clerk missed an item in your cart or you passed an unlocked bicycle on an empty street. In any of those situations, the SMORC thing to do would be to take the money, laptop, or bike or not mention the missed item. Yet we pass up the vast majority of these opportunities every day without thinking that we should take them. This means that we're off to a good start in our effort to improve our moral fiber.

What About "Real" Criminals?

Across all of our experiments we've tested thousands of people, and from time to time, we did see aggressive cheaters who keep as much money as possible. In the matrix experiment, for example, we have never seen anyone claim to solve eighteen or nineteen out of the twenty matrices. But once in a while, a participant claimed to have solved all twenty matrices correctly. These are the people who, having made a cost-benefit analysis, decided to get away with as much money as possible. Fortunately, we didn't encounter many of those folks, and because they seemed to be the exception and not the rule, we lost only a few hundred dollars to them. (Not exactly thrilling, but not too bad.) At the same time, we had thousands and thousands of participants who cheated by "just" a few matrices, but because there were so many of them, we lost thousands and thousands of dollars to them—much, much more than we lost to the aggressive cheaters.

I suspect that in terms of my financial losses to the aggressive and to the small cheaters, our experiments are indicative of dishonesty in society at large. Very few people steal to a maximal degree. But many good people cheat just a little here and there by rounding up their billable hours, claiming higher losses on their insurance claims, recommending unnecessary treatments, and so on. Companies also find many ways to cheat a little bit. Think about credit card companies that raise interest rates ever so slightly for no apparent reason and invent all kinds of hidden fees and penalties (which are often referred to, within companies, as "revenue enhancements"). Think about banks that slow down check processing so that they can hold on to our money for an extra day or two or charge exorbitant fees for overdraft protection and for using ATMs. All of this means that although it is obviously important to pay attention to flagrant misbehaviors, it is probably even more important to discourage the small and more ubiquitous forms of dishonesty—the misbehaviors that affect all of us most of the time—both as perpetrators and as victims.

A Word About Cultural Differences

I travel a lot, which means that I get to meet people from all over the world, and when I do, I often ask them about honesty and morality in their countries. As a result, I'm beginning to understand how cultural differences—whether regional, national, or corporate—contribute to dishonesty.

If you grew up outside the United States, think about this for a minute: do people from your home country cheat more or less than Americans do? After asking many people from various countries this question, I've discovered that people have very strong beliefs about cheating in their own countries, and most believe that people in their home country cheat more than Americans do (with the somewhat predictable exception of people from Canada and the Nordic countries).

Understanding that these are only subjective impressions, I was curious to see whether there really was something to them. So I decided to test some of these cultural perceptions more directly. In order to explore cultural differences, we first had to come up with a way to equate the financial incentives across the various locations. If we always paid, for example, an amount equivalent to $1 for a correctly solved question, this would range from being a very high payment in some places to a rather low one in others. Our first idea of how to equate the size of the incentives was to use a product that would be internationally recognized, such as a McDonald's hamburger. Following this approach, for each matrix solved correctly, participants could receive one-quarter of the cost of a McDonald's hamburger in that location. (This approach assumed that the people setting prices at McDonald's understand the economic buying power in each location and set their prices accordingly.)

In the end we decided on a related approach and used the "beer index." We set up shop in local bars and paid participants one-quarter of the cost of a pint of beer for every matrix that they claimed to have solved. (To make sure that our participants were sober, we only approached bargoers as they were entering the bar.)

BECAUSE I GREW up in Israel, I especially wanted to see how Israelis measured up (I admit that I suspected that Israelis would cheat more than Americans). But as it turned out, our Israeli participants cheated in the matrix experiments just as much as the Americans. We decided to check other nationalities, too. Shirley Wang, one of my Chinese collaborators, was convinced that Chinese people would cheat more than Americans. But again, the Chinese showed the same levels of dishonesty. Francesca Gino, from Italy, was positive that Italians would cheat the most. "Come to Italy, and we will show you what cheating is all about," she said in her fantastic accent. But she was proven wrong too. We discovered the same results in Turkey, Canada, and England. In fact, the amount of cheating seems to be equal in every country—at least in those we've tested so far.

How can we reconcile the fact that our experiments don't show any real differences in dishonesty among various countries and cultures with the very strong personal conviction that people from different countries cheat to different degrees? And how can we reconcile the lack of differences we see in our results with the clear differences in corruption levels among countries, cultures, and continents? I think that both perspectives are correct. Our data reflect an important and real aspect of cheating, but so do cultural differences. Here's why.

Our matrix test exists outside any cultural context. That is, it's not an ingrained part of any social or cultural environment. Therefore, it tests the basic human capacity to be morally flexible and reframe situations and actions in ways that reflect positively on ourselves. Our daily activities, on the other hand, are entwined in a complex cultural context. This cultural context can influence dishonesty in two main ways: it can take particular activities and transition them into and out of the moral domain, and it can change the magnitude of the fudge factor that is considered acceptable for any particular domain.

Take plagiarism, for example. At American universities, plagiarism is taken very seriously, but in other cultures it is viewed as a kind of poker game between the students and faculty. In those cultures getting caught, rather than the act of cheating itself, is viewed negatively. Similarly, in some societies, different kinds of cheating—not paying taxes, having an affair, downloading software illegally, and running red lights when there is no traffic around—are frowned upon, while in other societies the same activities are viewed as neutral or even confer bragging rights.

Of course, there's a great deal more to learn about the influence of culture on cheating, both in terms of the societal influences that help curb dishonesty and in terms of the social forces that make dishonesty and corruption more likely.

P.S. I SHOULD point out that throughout all of our cross-cultural experiments, there was one time we did find a difference. At some point Racheli Barkan and I carried out our experiment in a bar in Washington, D.C., where many congressional staffers gather. And we carried out the same experiment in a bar in New York City where many of the customers are Wall Street bankers. That was the one place where we found a cultural difference. Who do you think cheated more, the politicians or the bankers? I was certain that it was going to be the politicians, but our results showed the opposite: the bankers cheated about twice as much. (But before you begin suspecting your banker friends more and your politician friends less, you should take into account that the politicians we tested were junior politicians—mainly congressional staffers. So they had plenty of room for growth and development.)

CHEATING AND INFIDELITY

Of course, no book about cheating would be complete if it didn't contain something about adultery and the kinds of complex and intricate subterfuges that extramarital relationships inspire. After all, in the popular vernacular, cheating is practically synonymous with infidelity.

In fact, infidelity can be considered one of the main sources of the world's most dramatic entertainment. If modern-day adulterers such as Liz Taylor, Prince Charles, Tiger Woods, Brad Pitt, Eliot Spitzer, Arnold Schwarzenegger, and many others hadn't cheated on their spouses, the tabloid magazine and various entertainment news outlets would probably go belly-up (so to speak).

In terms of the fudge factor theory, infidelity is most likely the prototypical illustration of all the characteristics of dishonesty that we have been talking about. To start with, it is the poster child (or at least one of them) of a behavior that does not stem from a cost-benefit analysis. I also suspect that the tendency toward infidelity depends to a great extent on being able to justify it to ourselves. Starting with one small action (maybe a kiss) is another force that can lead to deeper kinds of involvement over time. Being away from the usual day-to-day routine, for example on a tour or a set, where the social rules are not as clear, can further enhance the ability to self-justify infidelity. And creative people, such as actors, artists, and politicians—all known for a tendency to be unfaithful—are likely to be more adept at spinning stories about why it's all right or even desirable for them to behave that way. And similar to other types of dishonesty, infidelity is influenced by the actions of those around us. Someone who has a lot of friends and family who have had affairs will likely be influenced by that exposure.

With all of this complexity, nuance, and social importance, you might wonder why there isn't a chapter in this book about infidelity and why this rather fascinating topic is relegated to one small section. The problem is data. I generally like to stick to conclusions I can draw from experiments and data. Conducting experiments on infidelity would be nearly impossible, and the data by their very nature are difficult to estimate. This means that for now we are left to speculate—and only speculate—about infidelity.

Figure 6:

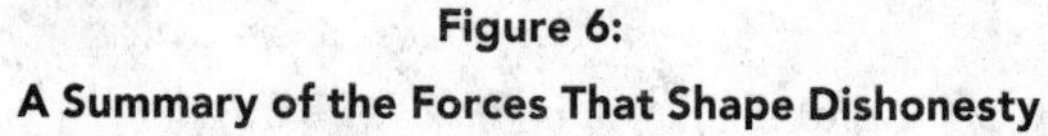

A Summary of the Forces That Shape Dishonesty

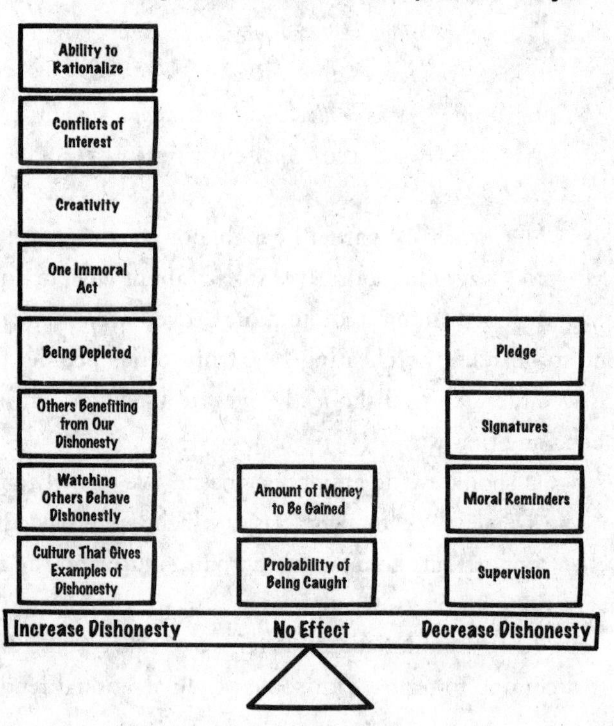

What Should We Do Next?

So here we are, surrounded by dishonesty. As one Apoth E. Cary put it in 1873:

Swindle, swindle, everywhere,
Every shape and size;
Take the swindle out of a man,
And you've nothing left but lies.
Philanthropy is made to cover a fraud,
Charity keeps humbugs in tow;
And we're swindled at home, swindled abroad,
And swindled wherever we go.
For the world is full of humbugs
Managed by dishonest men;
One moves on, another comes,
And we're swindled again and again.

—APOTH E. CARY, "RECOLLECTIONS OF THE SWINDLE FAMILY"[1]

As we have seen, we are all capable of cheating, and we are very adept at telling ourselves stories about why, in doing so, we are not dishonest or immoral. Even worse, we are prone to "catch" the cheating bug from other people, and once we start acting dishonestly, we are likely to continue misbehaving that way.

So what should we do about dishonesty? We recently experienced a tremendous financial crisis, which has provided an excellent opportunity to examine human failure and the role that irrationality plays in our lives and in society at large. In response to this man-made disaster, we've taken some steps toward coming to terms with some of our irrational tendencies, and we've begun reevaluating our approach to markets accordingly. The temple of rationality has been shaken, and with our improved understanding of irrationality we should be able to rethink and reinvent new kinds of structures that will ultimately help us avoid such crises in the future. If we don't do this, it will have been a wasted crisis.

MEMENTO MORI

There are a lot of possible connections one can draw between Roman times and modern-day banking, but perhaps the most important of them is *memento mori.* At the peak of Rome's power, Roman generals who had won significant victories marched through the middle of the city displaying their spoils. The marching generals wore purple-and-gold ceremonial robes, a crown of laurels, and red paint on their face as they were carried through the city on a throne. They were hailed, celebrated, and admired. But there was one more element to the ceremony: throughout the day a slave walked next to the general, and in order to prevent the victorious general from falling into hubris, the slave whispered repeatedly into his ear, *"Memento mori,"* which means "Remember your mortality."

If I were in charge of developing a modern version of the phrase, I would probably pick "Remember your fallibility" or maybe "Remember your irrationality." Whatever the phrase is, recognizing our shortcomings is a crucial first step on the path to making better decisions, creating better societies, and fixing our institutions.

THAT SAID, OUR next task is to try to figure out more effective and practical ways to combat dishonesty. Business schools include ethics classes in their curricula, companies make employees sit through seminars on the code of conduct, and governments have disclosure policies. Any casual observer of the state of dishonesty in the world will quickly realize that such measures don't get the job done. And the research presented here suggests that such Band-Aid approaches are doomed to fail for the very simple reason that they don't take into account the psychology of dishonesty. After all, every time policies or procedures are created to prevent cheating, they target a certain set of behaviors and motivations that need to change. And generally when interventions are set forth, they assume that the SMORC is at play. But as we have seen, this simple model has little to do with the driving forces behind cheating.

If we are really interested in curbing cheating, what interventions should we try? I hope it is clear by now that if we are to stand a chance of curbing dishonesty, we must start with an understanding of *why* people behave dishonestly in the first place. With this as a starting point, we can come up with more effective remedies. For example, based on our knowledge that people in general want to be honest but are also tempted to benefit from dishonesty, we could recommend reminders at the moment of temptation, which, as we've seen, are surprisingly effective. Similarly, understanding how conflicts of interest work and how deeply they influence us makes it clear that we need to avoid and regulate conflicts of interest to a much higher degree. We also need to understand the effects that the environment, as well as mental and physical depletion, plays in dishonesty. And of course, once we understand the social infectiousness of dishonesty, we could take a cue from the Broken Windows Theory to combat the social contagion of cheating.

INTERESTINGLY, WE ALREADY have many social mechanisms in place that seem to be designed specifically for resetting our moral compass and overcoming the "what-the-hell" effect. Such resetting rituals—ranging from the Catholic confession to Yom Kippur, and Ramadan to the weekly Sabbath—all present us with opportunities to collect ourselves, stop the deterioration, and turn a new page. (For the nonreligious, think of New Year's resolutions, birthdays, changes of job, and romantic breakups as "resetting" opportunities.) We have recently started carrying out basic experiments on the effectiveness of these types of resetting approaches (using a nonreligious version of the Catholic confession), and so far it seems that they can rather successfully reverse the what-the-hell effect.

From the social science perspective, religion has evolved in ways that can help society counteract potentially destructive tendencies, including the tendency to be dishonest. Religion and religious rituals remind people of their obligations to be moral in various ways; recall, for example, the Jewish man with the tzitzit from chapter 2 ("Fun with the Fudge Factor"). Muslims use beads called tasbih or misbaha on which they recount the ninety-nine names of God several times a day. There's also daily prayer and the confessional prayer ("Forgive me, Father, for I have sinned"), the practice of prayaschitta in Hinduism, and countless other religious reminders that work very much as the Ten Commandments did in our experiments.

To the extent that such approaches are useful, we might think about creating related (albeit nonreligious) mechanisms in business and politics. Maybe we should get our public servants and businesspeople to take an oath, use a code of ethics, or even ask for forgiveness from time to time. Perhaps such secular versions of repentance and appeal for forgiveness would help potential cheaters pay attention to their own actions, turn a new page, and by doing so increase their moral adherence.

ONE OF THE more intriguing forms of resetting ceremonies is the purification rituals that certain religious sects practice. One such group is Opus Dei, a secretive Catholic society, in which members flagellate themselves with cattail whips. I don't remember exactly how we started discussing Opus Dei, but, at some point Yoel Inbar (a professor at Tilburg University), David Pizarro and Tom Gilovich (both from Cornell University), and I wondered if self-flagellation and similar behaviors capture a basic human desire for self-cleansing. Can the feeling of having done something wrong be erased by self-punishment? Can self-inflicted pain help us ask for forgiveness and start anew?

Following the physically painful approach of Opus Dei, we decided to conduct an experiment using a more modern and less bloody version of cattail whips—so we picked mildly painful electric shocks as our experimental material. Once participants came to the lab at Cornell University, we asked some of them to write about a past experience that made them feel guilty; we asked other participants to write about a past experience that made them feel sad (a negative emotion but not related to guilt); and we asked a third group to write about an experience that made them feel neither good nor bad. After they reflected on one of these three types of experiences, we asked the participants to take part in "another" experiment involving self-administered electrical shocks.

In this next phase of the experiment, we connected the participant's wrist to a shock-generating machine. Once the connection was secure, we showed the participants how to set the level of the electrical shock and which button to press to give themselves the painful jolt. We set the machine to the lowest possible level of shock and asked participants to press the switch, increase the level of the shock, press the switch, increase the level of the shock, press the switch, and so on until they could no longer tolerate the intensity of the shock.

We really aren't as sadistic as it might sound, but we wanted to see how far participants would push themselves on the pain scale and to what extent their level of self-administered pain would depend on the experimental condition they were in. Most important, we wanted to see whether being reminded of a guilt-related past experience would cause our participants to cleanse themselves by seeking more pain. As it turned out, in the neutral and sad conditions, the degree of self-inflicted pain was similar and rather low, which means that negative emotions by themselves do not create a desire for self-inflicted pain. However, those in the guilty condition were far more disposed to self-administering higher levels of shocks.

As difficult as it might be to appreciate this experimental support for the practice of Opus Dei, the results suggest that purification through the pain of self-flagellation might tap into a basic way we deal with feelings of guilt. Perhaps recognizing our mistakes, admitting them, and adding some form of physical punishment is a good recipe for asking forgiveness and opening a new page. Now, I am not recommending that we adopt this approach just yet, but I can think of some politicians and businessmen whom I would not mind trying it out on—just to see if it works.

A MORE SECULAR (and more elegant) example of resetting was told to me by a woman I met at a conference a few years ago. The woman's sister lived in South America, and one day the sister realized that her maid had been stealing a little bit of meat from the freezer every few days. The sister didn't mind too much (other than the fact that sometimes she didn't have enough meat to make dinner, which became rather frustrating), but she clearly needed to do something about it. The first part of her solution was to put a lock on the freezer. Then the sister told her maid that she suspected that some of the people who were working at the house from time to time had been taking some meat from the freezer, so she wanted only the two of them to have keys. She also gave her maid a small financial promotion for the added responsibility. With the new role, the new rules, and the added control, the stealing ceased.

I think this approach worked for a number of reasons. I suspect that the maid's habit of stealing developed much like the cheating we've been discussing. Perhaps it began with a single small action ("I'll just take a little bit of meat while I'm cleaning up"), but having stolen once, it became much easier to continue doing so. By locking the freezer and giving the maid an additional responsibility, the sister offered the maid a way to reset her honesty level. I also think that trusting the maid with the key was an important element in changing her view on stealing meat and in establishing the social norm of honesty in that household. On top of that, now that a key was needed to open the freezer, any act of stealing would have to be more deliberate, more intentional, and far more difficult to self-justify. That is not unlike what happened when we forced participants to deliberately move the mouse to the bottom of the computer screen to reveal an answer key (as we saw in chapter 6, "Cheating Ourselves").

The point is that the more we develop and adopt such mechanisms, the more we will be able to curb dishonesty. It is not always going to be simple, but it is possible.

IT'S IMPORTANT TO note that creating an endpoint and the opportunity for a new beginning can take place on a broader social scale. The Truth and Reconciliation Commission in South Africa is an example of this kind of process. The purpose of this courtlike commission was to enable the transition from the apartheid government, which had sharply oppressed the vast majority of South Africans for decades, to a new beginning and to democracy. Similar to other methods of stopping negative behavior, pausing, and starting again, the goal of the commission was reconciliation, not retribution. I'm sure that no one would claim that the commission erased all memories and remnants of the apartheid era or that anything as deeply scarring as apartheid could ever be forgotten or fully healed. But it remains an important example of how acknowledging bad behavior and asking for forgiveness can be an important step in the right direction.

FINALLY, IT IS worth trying to examine what we have learned about dishonesty from a broader perspective and see what it can teach us about rationality and irrationality more generally. Through the different chapters, we have seen that there are rational forces that we think drive our dishonest behavior—but don't. And there are irrational forces that we think don't drive our dishonest behavior—but do. This inability to recognize which forces are at work and which are irrelevant is something we consistently see in decision making and behavioral economics research.

Viewed from this perspective, dishonesty is a prime example of our irrational tendencies. It's pervasive; we don't instinctively understand how it works its magic on us; and, most important, we don't see it in ourselves.

The good news in all of this is that we are not helpless in the face of our human foibles (dishonesty included). Once we better understand what really causes our less-than-optimal behavior, we can start to discover ways to control our behavior and improve our outcomes. That is the real goal of social science, and I am sure that the journey will only become more important and interesting in the years to come.

Irrationally yours,
Dan Ariely

CHAPTER 11

Some Reflections on Religion and (Dis)honesty

For the past few years, whenever I ate at a restaurant I'd ask the waiters if there was a way to eat and leave without paying. Sometimes the waiters would ask for my credit card as security, but most times, they'd give me good suggestions. For example, they'd tell me I could go to the bathroom and use the side door to leave before the bill came. Or they'd tell me that I could get out my wallet and pretend to pay in cash, then nonchalantly walk away from the table.

Armed with this advice about how to dine-and-dash, I'd ask the waiters how often people actually did this. And without exception, they would tell me that it almost never happened, despite the chance for a free meal and the very high likelihood of getting away with it. This, and other behaviors like it, shows that while it would be selfishly beneficial for us to dine-and-dash, we operate within a set of moral constraints that makes us bypass this kind of immoral behavior.

In contrast, let's turn to the case of illegal music downloads. Sometimes I ask my students if they have any illegal music on their computers, and almost all of them admit that they do—and that they don't care if their friends or parents find out about it. So what's the difference between illegal downloads and skipping out on the restaurant bill? It's clearly not a matter of being caught, as the chances for that in both cases are very low. It is also not about the magnitude of punishment, because if someone caught you not paying your bill at the restaurant you could always say it slipped your mind, while the music industry would certainly not buy the excuse "I just forgot, sorry." To a large degree, what stops us from behaving badly is not the probability of being caught or the size of the punishment. Instead, it is the guilt that comes from having a direct connection with the restaurant staff, which basically activates our internal monitor and our consciences.

The good news is that we all have a moral compass. The bad news is that we can't just assume that our consciences effortlessly and continuously protect us. So how do we keep our moral compass in good working order? For answers, let's look at some ancient wisdom.

Some Insights from Religion

You probably noticed that I've referred to examples from religion quite often in the preceding pages. I mentioned religious traditions in discussing the experiment on the effect of remembering (or trying to remember) the Ten Commandments on dishonesty; in the story about the man whose tzitzit prompted him to forgo unsanctioned carnal pleasure (p. 45); in the role of resetting rituals for countering the what-the-hell effect (p. 249); and in the connection between Opus

Dei's practice of self-flagellation and our guilt-primed experimental participants (p. 251). Why did I do that?

The answer is this: religion has a lot to say about our struggle with a range of human problems, including honesty and morality. Of course, my choice to look to religion for ideas could also be a reflection of my maturity and age (people sometimes become more spiritual as they grow older). Regardless of the reason, and regardless of any belief in God, looking at these texts more broadly as a reflection of human thought and wisdom could help us shed some light on honesty, dishonesty, and their complex place in society.

A Brief Proviso . . .

So what are the potential lessons from religion on dishonesty? Before diving in, I should mention that when I describe stories from Scripture, I am aware there are different versions and nuances. I am not an expert in religion; my understanding of these topics is limited in general, and further limited to my knowledge of Judaism and a bit of Christianity. I was informed of my limitations in this domain innumerable times following the publication of *Predictably Irrational*, where I listed the Roman Catholic version of the Ten Commandments and forgot to mention that there are other versions (not to mention moral codes from other religions). So if the religious stories and lessons I describe in the following pages aren't the ones you're used to, I hope you can overlook the differences and still find the lessons helpful. And I ask for the patience and forgiveness of all non-Jews and non-Christians in advance.

Social Science and Religion

Seen through the lens of social science, we can think of religions not only as sets of particular beliefs about God, but also as prescriptive sets of insights and rules for behavior. From this perspective, religious principles have an important role in directing people to behave in ways that help us better coexist with one another, and with a view toward long-term goals rather than short-term self-interest.*

As initial evidence that ancient religious texts have something to say about the topics that social scientists struggle with, here is an example of how religious thought deals with one of the major differences between social psychology and neoclassical economics: Do our preferences follow our actions, or do our actions follow our preferences? Economists say that our preferences drive our behaviors. We start with what we like, and from those stable preferences we make decisions and purchases. On the other hand, psychologists (and behavioral economists) say that the direction of causality is sometimes reversed—that is, our actions can drive (or at least influence) our preferences. For example, when we build something, the act of building it causes us to like it even more.

And what do religions say about this question? In Sanhedrin 105b, a Jewish rabbinical text, there is deep appreciation of the self-signaling feedback loop: "A person should always engage in Torah and its precepts even if not with sincerity, because insincere behavior leads to sincere behavior." A me-

* Of course, over the ages people have done tremendous harm and caused bloodshed in the name of their religions, but for our purposes here, I would like to focus on the way religion has dealt with honesty and dishonesty and on the positive lessons we can take from it.

dieval rabbi, Aharon Halevi of Barcelona, took this idea even further:*

> Know that a person is influenced by his actions, and his heart and all his thoughts always follow the acts he does, whether they are good or bad. Even one who is a completely wicked person, whose heart conspires bad all the day, if his spirit is aroused and he invests his efforts and occupation in diligently fulfilling Torah and mitzvot, even if it is not for the sake of Heaven, he will automatically turn to the good and by the power of his deeds he will crush his Evil Inclination, since the heart follows the actions a person does. Similarly, even if one is completely righteous and his heart is straight and sincere, and he desires Torah and mitzvot but always involves himself in deeds of vanity, for example if the king forced him and appointed him to an evil trade, truly if his entire occupation is in that trade, after a while he will be transformed from an honest heart to a wicked person, for it is known and it is true, that every man is affected by his actions.

I should point out that I was particularly delighted that behavioral economics and the role of behavior in forming preferences (a process we have referred to as self-signaling) emerge as the prevailing view here.

AS AN EXPERIMENTALIST, I was also pleased to find that the best method I know for finding out what is going on—conducting experiments—is documented in the Bible. For example, in Judges 6, Gideon tries to answer the question, "How can I know for sure that something's true—that I'm

---

* Mitzavh 16 in *Sefer Hachinuch* (attributed to Rabbi Aharon Halevi of Barcelona, thirteenth century).

not just believing what I want to believe?"* Gideon's experiment was to test whether it was indeed God who spoke to him and wanted him to lead a rebellion, or whether it was just a voice in his head. On the first night Gideon was contemplating this question, he asked God to make the morning dew land only on a piece of wool fleece and not on the ground around it. The next night, he wanted to make sure that this dew-falling pattern of data was not due to chance or particular weather conditions. So he set a control condition, and asked God to do the exact opposite: keep the fleece dry, but surround it with dewy ground. When he saw both patterns of data confirmed, he was satisfied that God was indeed on his side, and the rebellion started.

Lessons about Conflicts of Interest

The first chapter of the book of Daniel is chock-full of social science. It's not what you expect to see as the introduction to a book known more for symbolic dreams and prophecies.

The book of Daniel begins in the sixth century BCE, when the Babylonian king Nebuchadnezzar was building his enormous Asian empire and looking for ways to reduce the risk of revolts among the peoples he'd conquered. Like a new CEO following a merger, he exiled whole groups of people. At the same time, he searched for ways of getting the people who lived at the edge of the empire to assimilate. In the case of the Israelites, he put together an incentive-based strategy.

After he occupied parts of the Holy Land, Nebuchadnezzar decided to create a conflict of interest among the Jewish

* See also *The Upside of Irrationality* (chapter 11).

elite. He took the handsomest and brightest of the newly conquered Israelite youth (including Daniel) to the palace and used incentives to shift their loyalties to him. He gave the boys royal food and placed them under the direct care of the king's chief of staff. They studied Babylonian language and literature as part of a three-year training course. At the end of it, Israel's cream of the crop would be set to become Babylonian officials, working for the empire as authorities, rather than slave laborers.

Giving people favors is a time-honored way of gaining loyalty. Pharmaceutical sales reps do it. The salespeople manning cosmetic counters do it. Lobbyists do it. Men with big crushes on impossibly beautiful women do it. Gifts work on our feelings in a couple of ways: they change the way we experience something, and they push our "reciprocate!" button. As you recall from Chapter 3, people preferred art from the gallery that was indirectly paying them. When we have a mandate to be objective and an incentive not to be, our biases often win the day—even if we don't think they will. Favors deeply affect our preferences and our loyalties.

So if the young Jewish men in the Babylonian palace were anything like the rest of us, Nebuchadnezzar's gifts of food and wine—not to mention prestige—might have softened the Israelite boys' stance on his invasion of the Holy Land. And it would have happened without the boys' even knowing how their feelings and loyalty had been altered. They might simply have attributed their changes in attitude to their own rational judgment and experience with the apparently gracious Chaldeans.

But beyond the bias Nebuchadnezzar's good treatment might have created in the boys themselves, the emperor was probably also hoping to influence their families. Maybe Nebuchadnezzar wanted the boys' families and the other exiles to be impressed by all the glamour and recognition showered upon their favorite sons. If these families saw that the king had given their loved ones favors, they might take on a bias themselves. Just imagine how difficult it would be to hate a person who is exceptionally nice and kind to our kids, especially if our kids love him back.

Interestingly, not everyone fell for Nebuchadnezzar's approach. Handsome Daniel and three of his friends (Hananiah, Mishael, and Azariah) found themselves at the top of their training class, but there's no indication that they did the ancient equivalent of golf outings or deep-sea fishing with the boss. Feeling that they owed their allegiance to the Israelites, they decided not to eat the meat and drink the wine the king sent them. As the four young men refused to take the king's food and wine, they were also refusing to take on the conflict-of-interest bias toward the king. And rather than let their loyalties inch toward Babylon (and, within a generation, Persia), they maintained their loyalty to the Jewish people and their religion. At the end of the three-year training period, the four of them were still rejecting the food and wine from Nebuchadnezzar.

When it was clear that Daniel would be the ideal choice for a promotion to the highest office in the Medo-Persian Empire, his enemies passed a law saying that for thirty days, no one could worship or pray to anyone except to the king. Daniel, of course, did not comply. His enemies were easily able to catch him praying to his God, and as a consequence he was thrown into a den of hungry lions. But, as the story goes, he miraculously survived and emerged from the lion den unscathed. Perhaps the story is a lesson for all of us who are tempted from time to time with conflicts of interest.

THE STORY OF Daniel and his friends shows us how to resist conflicts of interest: try as hard as you can not to accept gifts that could sway your judgment. If you must accept a dinner on someone else's tab, realize that it might be an effort to change your thinking and behavior. And if you're in a situation in which the conflict of interest is unavoidable, decide in advance to set up a clear ranking of loyalties and priorities and make sure that you remember and follow them.

Many religions have explicit rules to deal with this exact problem. The path to ownership involves social and business transactions, and throughout the process of acquiring positions and belongings, social obligations can spring up in the forms of biases and divided loyalties. Because religious leaders are in positions of power to start with, to quash conflicts of interest entirely, many leaders of Catholic, Buddhist, and other orders have to give up the right to own anything when they join.

Hundreds of years before Daniel was born, Moses warned that conflicts of interest can change a person's perspective: "Do not accept a bribe, for a bribe blinds those who see and twists the words of the innocent." For all of us who lived through the financial meltdown of 2008, it's easy to see why conflicts of interest would be a public policy issue. Large sums of money skewed the objectivity of people who should have recognized the danger posed by sketchy financial tools. And even after all the damage was done and we gained a new understanding of the harmful forces in the financial markets, lobbying still effectively prevents new and better regulations from being put in place.

White Lies in a Gray World

For me, perhaps the most important (and probably least acknowledged) lesson from religion that pertains to this book is the recognition that some level of dishonesty is actually needed in society. You've heard it before: life is not black and white. When we're making decisions, we're looking at a lot of pros and cons all jumbled together into a spectrum of gray. Our motivations—no matter how honorable—often counteract other motivations.

ABRAHAM AND SARAH provide a wonderful example of this in Genesis 18:1–14 (New International Version). This fine Biblical couple didn't have any children, despite the fact that God had told Abraham multiple times that he and Sarah would have more offspring than they could ever count. And God reaffirmed his promise even after they were past childbearing age.

Sarah had been eavesdropping when God spoke with Abraham, and chuckled to herself when she heard God's words, thinking how unlikely it was that she could have a son when Abraham was so old. And she was no spring chicken, either.

God overheard Sarah's laughter and then offered Abraham a slightly modified explanation of her reaction: "Then the Lord said to Abraham, 'Why did Sarah laugh and say, "Will I really have a child, now that I am old?" Is anything too hard for the Lord? I will return to you at the appointed time next year, and Sarah will have a son.'"

Some religious scholars have an interesting interpretation of this passage. Rather than focusing on Sarah's unbelief in the power of God, they point out that God was protecting Abraham from learning what Sarah was really thinking: that Abraham was too old to, well, produce. In explaining Sarah's laughter to Abraham, God seems to have left out the part where Sarah had a good chuckle about Abraham's age and virility. Instead, God made it sound like she didn't believe in his powers. In short, God told a fib. Why?

Rashi, the famous medieval rabbi, thought that this passage illustrated that there's such a thing as too much honesty. Sometimes, complete honesty puts domestic peace at risk. Much like God in this passage, we too might need to prioritize familial harmony over complete disclosure from time to time (which is an amazing rationalization, if you ever need one).

I talked about this interpretation with the chief rabbi of England, Jonathan Sacks, and he agreed with Rashi. Rabbi Sacks told me that the story illustrates the need to prioritize multiple values when they're in conflict. "There are many human values," he said. "Honesty is one of them. Peace in the family is another. But sadly, not all human values are compatible at all times in all circumstances. What happens when values collide? In Judaism, the teaching is that peace at home can sometimes trump full honesty." (I should also point out that Rabbi Sacks added that this does not mean we should take dishonesty lightly, even when we're telling a white lie, and even if it is for peace at home.)

In contrast, the eighteenth-century philosopher Immanuel Kant presented a very different perspective on dishonesty. Kant famously put forward the idea that one should *never* compromise when it comes to honesty. Kant believed that honesty was a mark of rationality, and that rationality was the foundation of human dignity. One critique challenges Kant's premise with the following scenario: Imagine that somebody wants to murder your friend, and you've hidden your friend in your house. The would-be murderer asks you whether your friend is hiding in your house. Even then, Kant says, you should tell the truth.*

However, the reality is that almost all of us—aside from Immanuel Kant and a few like him—recognize that a bit of dishonesty provides valuable social cushioning. Almost no one thinks the best answer to the age-old question "Honey, how do I look in this dress?" or the proper response to a do-it-yourself home project is always the bare truth. If we are honest with ourselves about our own dishonesty, we must recognize that in the domain of social relationships we often tell white lies, and that we don't want to live with people who tell us the whole truth all the time. In fact, we want the people around us to fudge things a bit in our favor. After all, without such social niceties, our relationships would fray and wear out rather quickly. White lies are useful for helping us preserve social relationships, and they are one of the main reasons we have such a complex relationship with the truth.

* From "On a Supposed Right to Lie Because of Philanthropic Concerns."

Another Look at the Ten Commandments: Keeping Institutions Honest

I had another interesting realization when I looked more closely at the eighth commandment* (Exodus 20:16), which states that people shall not bear false witness. In my memory it was referring to lying. But it is not. In fact it is about perjury. Perjury?

Let's face it: if we each had to come up with ten things we didn't want anyone to ever do, most of us would not include giving false testimony. Nevertheless, a prohibition on perjury made it into the Ten Commandments, and the matter does not end there. Later on, in Deuteronomy 19:15–21, there is a description of a scene in which the judges start to doubt a witness. In such a case, the text instructs, the judges should make a thorough investigation of the witness's testimony, and if they find the witness to be lying: "then do to the false witness as that witness intended to do to the other party. You must purge the evil from among you. The rest of the people will hear of this and be afraid, and never again will such an evil thing be done among you."

When I first looked at this commandment, it seemed not to tackle the real issue of lying and dishonesty. I also thought it seemed unnecessarily specific. I suspect that most people would agree, after thinking about this for a moment, that perjury is not on the same level as murder or theft. Does anyone's toes curl at the idea of lying in court? Why is false witness given such priority—both in the Ten Commandments and in

* Depending on the religion or denomination, this is sometimes the ninth commandment.

the modern legal system that severely punishes perjury—among all of our potential dishonest acts?

One simple explanation is that false testimony can have a devastating effect on the innocent. People could be punished and even killed because of a false witness ("Yes, Your Honor, it was my ex-wife that stabbed this innocent person on the street the other day"). But if this was the main reason for the ban on false testimony, the same outcome could be achieved with a commandment that forbids lying more generally.

If we think a bit more about the unique nature of false witness, it turns out that lying in court is grievously problematic on a number of levels. First, any lying in court is a public act. And as we showed in the "bad apple" experiment in Chapter 8, once even one person obviously and egregiously cheats, more people from the same social group start finding this kind of behavior to be more socially acceptable. This is why public dishonesty in the form of bearing false witness can quickly elevate dishonesty to new heights—and why it is more important to prevent than our private dishonesty.

A second issue (and the one I'm most interested in for our purposes here) is that lying in court interferes with good governance. Disabling institutions may not have the same shock value as committing violent crimes, but it does erode the trust we all have—and need to have—in our public institutions if we want them to function well. (If a CEO of a large bank gets away with no penalty after betraying his customers and shareholders, why shouldn't I pretend that I have no idea how the TV broke in my warranty claim? And why should I pay my taxes? Etc.) This is why the consequences of such crimes against social institutions can be devastating, potentially eroding the rule of law and order in society.

For all of these reasons (the effect on the innocent, the public erosion of honesty, and the importance of good institutions), I suspect that the commandment against false witness is actually very important. And while this commandment does not provide a general prohibition on lying, its focus is important for the health of society.

The Importance of Rules

During one of our conversations, I asked Rabbi Sacks to tell me which of the Ten Commandments I should keep, if I were going to focus on just one. It was another way of asking him which commandment is the most important one. What do you think his answer was? The one about not worshipping idols? The one about murder?

HIS ANSWER WASN'T at all what I expected—he said that if I kept only one commandment, I should observe Shabbat. "If you keep Shabbat as a day of rest and reflection," he said, "the rest of the commandments will most likely follow." Apparently he had come to hold the same general idea as social scientists about the way our moral muscles and ego depletion work—and about the importance of rest and resetting our moral energy (see Chapter 4).

The Bible begins by describing how God created the world. It goes day by day. He started by creating light, then he formed the earth, then he created sea creatures and plants, and so on, up to the point when he made land animals and people. After all of that, God reserved the seventh day for something quite different: rest. It was the first Shabbat.

The Shabbats after that were for people and for the land. God basically said that every seventh day would be a day of rest, when people would not gather food, buy, sell, farm, or carry loads. Every seventh year, farmland would be left fallow. Observing Shabbat became one of the Ten Commandments, which God engraved onto stone tablets and gave to Moses: "Remember the Sabbath day by keeping it holy. Six days you shall labor and do all your work, but the seventh day is a Shabbat to the Lord your God."

Shabbat affects those who observe it in a few ways. First, it offers an opportunity to stop and reflect. In observing this day, we can remind ourselves what we have done in the last week, what we want to do next week, and what our true values are. We can pay attention to our less-than-perfect behaviors that otherwise might go unnoticed, keeping ourselves from sliding accidentally into moral dangers.

The second way Shabbat propels people to observe the other commandments is by restoring our moral energy. It's no secret that at the end of a day or week, people often let loose (getting drunk and so on) by allowing themselves to do what their impulsive id-side has been screaming for while they were stuck in their cubicles. We saw this kind of moral exhaustion in the depletion experiment with Nicole Mead, Roy Baumeister, Francesca Gino, and Maurice Schweitzer (see Chapter 4). In that experiment, participant cheating increased after a more difficult writing task, suggesting that in daily life (even without hard writing assignments from social scientists) exhaustion can wear us right down to our ids. Ego depletion (as we call this draining effect), it turns out, affects not only whether we make good or bad decisions, but also whether we obey our consciences.

There's another way that Shabbat might be a boon to our morality. It has to do with rules and self-control. Decisions are among the many behaviors that wear us down. Decisions make life more complex and difficult. The more of them we have to make, the weaker our self-control becomes. What can relieve people from making decisions? Rules. Shabbat isn't just about relaxing. It's also about keeping a list of rules and restrictions—things that make the day different from all others. Rules offer relief, because somebody else (in this case, God) is doing the driving, and you can doze off in the passenger seat.

If you haven't celebrated an Orthodox Jewish Shabbat, you might not be aware of all the rules. Orthodox Jews don't drive, ride in cars, bake, write, engage in any commercial activity, ignite fires, or put fires out. Very observant Orthodox Jews won't even rip anything on Shabbat, so they use tissues instead of toilet paper.

Of course, following these rules can take some planning ahead. But it also dramatically simplifies decisions about what to wear and what to eat, whether to work, watch TV, answer e-mail, or spend time with the kids, and so on. One twentieth-century rabbi, Eliyahu Dessler, suggested that rules (or "decision points," as he called them) serve as swords and shields in the moral battleground. Habitual Shabbat observers, he would say, aren't expending any moral energy deciding what to do during the Sabbath or whether to break the rules of Shabbat.* These decisions have already been made by God, and now all we have to do is execute the deci-

* Rabbi Eliyahu Eliezer Dessler, *Letter from Eliyahu* (Bnei Brak: Sifriati [Gitlers] LTD, 2002).

sions without question and doubt—and without the associated mental anguish and exhaustion.

But can we really reduce ego depletion? And does following rules really help us get a handle on temptation? One very interesting study by Reuven Dar, Florencia Stronguin, Roni Marouani, Meir Krupsky, and Hanan Frenk* tried to answer this question. They approached Orthodox Jews who were heavy smokers and asked them to report their moods and cigarette cravings as they went about their day. The researchers repeated this procedure on three different types of days: 1) a regular day when the religious smokers smoked as usual; 2) a regular day when they could smoke, but the researchers asked them to refrain from smoking (and the religious smokers complied); and 3) Shabbat, when the religious smokers were not allowed to smoke, according to the rules of their faith. The question was whether strict Shabbat rules would reduce the smokers' feelings of temptation and need for self-control.

So let's say you are one of the smokers in this study. On which day would you have the most craving, and on which day would you have the least craving? If the only factor that determined your level of craving is how recently you had smoked a cigarette, you should experience the same mood and craving patterns on the two days when you didn't smoke (the regular day without smoking and the Sabbath). And we would expect you to crave cigarettes the least on the regular day when you did smoke. The question is, what if you're an Orthodox Jew? Would strict

* "Craving to smoke in orthodox Jewish smokers who abstain on the Sabbath: a comparison to a baseline and a forced abstinence workday." Reuven Dar, et. al. *Psychopharmacology* (2005) 183: 294–299.

Shabbat rules mean less temptation and lower need for self-control on the day you did not smoke because of religious reasons?

And the results? The participants reported a substantial difference between their cravings on Shabbat and on the two other days. On Shabbat, they reported much lower degrees of craving relative to the regular day without smoking. They also said they didn't even crave cigarettes as much as they did on the regular day when they did smoke. Interestingly, it was the Sabbath, and not smoking, that reduced craving the most!

Of course, we should always be suspicious of self-reported data, especially when the questions are related to violating social or moral rules. But the main finding (that smokers said they craved cigarettes much *less* on Shabbat) is nevertheless interesting.

These findings and ideas suggest that rules, which are the foundation of many of the world's religions, can help us deal with temptation. They can help to "lead us not into temptation," as the Lord's Prayer says, "but deliver us from evil." And they help not only with the decision at hand but also with future decisions. When we have to make lots of decisions with no rules to guide us, nothing is automatic and we have to invest some energy in every one of those decisions. All of those decisions sap our moral energy, making us more likely to yield to temptation. But when we have rules—particularly strict rules—we don't have to make the same number of decisions, and we are left with much more moral energy. Of course, this same principle applies to all rules, religious and nonreligious, as long as they are clear and help reduce the burden of choice from our lives.

I am not sure about you, but while I am not particularly looking forward to having more strict rules in my life, I do find it intuitively appealing that rules (like the Golden Rule) can help us regulate our behavior. When we think about what rules we should add to our lives, we should ask ourselves what the characteristics of useful rules are. The first characteristic for rules seems to be that they have to give us a specific action plan. For example, in Alcoholics Anonymous (AA), the specific action plan is to set a time period in which one will not drink alcohol—an hour, a day, a week, a month, and so on. The idea is to build a series of small, concrete steps.

The second characteristic seems to be that the rules have to be precise, so that it is easy for us to see at any moment which side of the rule we are on. Think, for example, about what would happen if AA members were allowed to drink half a glass of alcohol a day. What size glass would AA members start buying and using? And what would happen if they started borrowing from their future supply, drinking more today while promising to drink less next week? Given such imprecise rules, members would quickly find themselves in a no-benefit situation. In contrast, the very clear rule of *no alcohol whatsoever* makes it easy to figure out whether you're keeping or breaking the rule.

Finally, I suspect that one other important characteristic for making rules is to have each rule link to a larger meaning. If the rule is set in an arbitrary way (exercise for thirty minutes, three times a week; eat two pieces of fruit and up to two thousand calories a day), the rule itself, and breaking it, is going to be relatively meaningless. But if the rules link us to other people (we are all doing this together), to some other larger purpose (this is what good people do), or to a deep belief (God's commandments), breaking the rule is more difficult and less likely to happen. In AA, for example, everything is linked to a sense of surrender to a "higher power."

So what if you had a specific and precise rule that on Saturdays or Sundays you shouldn't use any digital devices or smoke, and what if you believed that keeping this rule served a higher purpose? With an absolute ban on those activities, you might find it easier to adhere to the plan. In addition, once you got past the point when you had to make active decisions about those activities, you would be in habit territory, and your decisions would be executed effortlessly. With these changes under your belt, you'd likely be able to resist other temptations—overeating, texting and driving, etc. If the logic holds, any version of a Sabbath might help you be your best self, even after sunset. I am not yet sure that this will always be the case, but it is certainly worth trying out.

On Keeping One Another Honest

What do good cooking, fake arm hair, and bad parents have in common?

No, the answer doesn't have to do with reality TV. These three things were among the tools Jacob and his mother, Rebekah, used to trick the rest of their family into making Jacob the heir.

Jacob's family was in a kind of arms race of deception and revenge. Isaac, Rebekah, and their two sons didn't just lie to one another; they came up with elaborate schemes involving impersonation, and they specialized in taking advantage of one another when emotions were running high.

Most often when we see cases of dishonesty, selfishness is one of the driving forces. But Jacob and Rebekah's story is a bit different. It's a case in which one is dishonest for the benefit of a loved one.

Isaac, the son of Abraham and Sarah, and his wife, Rebekah, had twin sons. Esau was born first, and he was one weird baby. The Bible describes him as red-faced, with a hairy body. His twin, Jacob, was born grasping Esau's heel. Even though there was a difference of only a few minutes between them, tradition dictated that the oldest son would inherit pretty much everything, including a special blessing from his father. Oh, and the parents played favorites—Isaac preferred Esau (the Bible hints that it was because Isaac liked Esau's cooking so much), and Rebekah preferred the underdog, Jacob.

Over the years, hairy baby Esau became just plain hairy Esau, and he married a couple of Hittite women whom his parents hated. Isaac was getting very old, and unless something changed, Isaac would irrevocably pass his property and blessing on to Esau and his descendants. So Jacob and Rebekah took advantage of Esau's long hunting trips to trick Isaac into giving Jacob Esau's inheritance.

When Esau got back to the tents after one of his trips, Jacob was cooking some tasty red bean soup. Esau thought he would die of starvation if he didn't get some of Jacob's food immediately. Jacob made an unorthodox suggestion: he would take Esau's inheritance in exchange for a bowl of red bean soup. Esau was either dumb as a rock or in a state of severe ego depletion, because he fell for it! "Look, I am about to die," Esau said. "What good is the birthright to me?" His hunger and exhaustion had drained him of all self-control.

But the story doesn't stop there. As he thought he was dying, Isaac—who was blind by then—wanted to give Esau a special, prophetic blessing. He called Esau in for a private conversation that Rebekah somehow managed to overhear. Isaac told Esau to make him a special meal and bring it with him when he came to receive his blessing. Esau went off to hunt for something to cook. In his absence, Rebekah quickly went to work. She prepared the meal she knew Isaac had asked Esau to make. Then she took hairy goatskins and strapped them onto Jacob's arms and neck so that Isaac would think he was Esau. Then she sent Jacob, in disguise, to Isaac to steal Esau's blessing. It worked.

When Esau found out that Jacob had tricked him out of the blessing as well as his birthright, he was angry enough to want to kill Jacob. But Rebekah already had a plan for getting Jacob to safety so he could preserve his victory. Before Esau came back, she told Isaac that Esau's wives were driving her crazy. "If Jacob takes a wife from among the women of this land, from Hittite women like these, my life will not be worth living," she told her husband. So Isaac sent Jacob off to find a wife far, far away. The journey was long enough for Esau to get over his initial anger and lighten up a bit—exactly according to Rebekah's scheme.

Leaving aside judgment about this family's clear dysfunction, what do you think of the question of parental neutrality here? Aren't parents supposed to make sure their children, regardless of birth order, do the right thing? In this story, Rebekah, in trying to help out her favorite son, manages to flip Jacob and Esau's seniority completely. Why would she do that? What could she have gained?

We often think of parents as supervisors, forgetting the emotional obligation they feel toward their children and vice versa. For Rebekah, there was likely a tension between her role as a parent who raised Jacob and Esau and her desire to help Jacob, who she felt was treated unfairly. But if you remember from Chapter 5, when we added a social element to the mix of forces that float around us there was an increase in dishonesty. "Social utility" is one good answer to the mystery of Rebekah's behavior. Social utility is social scientists' way of describing our general urge to take care of others, even when doing so uses up our own resources. There's no doubt that taking social utility into account often makes our world better (think of all the charities, good deeds, and programs that improve lives). But at the same time, it can also justify cheating, lying, or worse for someone else's sake.

Think back to Chapter 9, in which Francesca Gino, Shahar Ayal, and I looked at altruistic cheating. We found that knowing that others will benefit from our actions does indeed motivate people to cheat more. Many people, from famous religious leaders to people with little influence or power, have lied in the name of their religion. Even if they realize that they're doing something wrong, they excuse their behavior by telling themselves that it is for the good of other people. Altruism is a powerful rationalization. In the same way, Rebekah had normal human reasons—if not rational ones—to help her favorite son. In any case, their story is a fine example of collaborative cheating.

A Final Reflection

There are many stories in the Bible that relate to the questions raised in this book. There is the story of Zacchaeus and how forgiveness gave him the opportunity to start fresh (related to Chapter 1); the question of the fudge factor and Abraham's not-so-white lies (Chapter 2); of how Daniel and his friends resolved their conflict of interest (Chapter 3); of how the Sabbath helps us keep other commandments (Chapter 4); of Rahab the prostitute possibly changing the course of her life with one act of goodness that led to other kind acts (Chapter 5); of self-deception and idolatry (Chapter 6); of King Solomon's creativity and social connections leading him to cheat more and more (Chapters 7 and 8); of Jacob and Rebekah's collaborative cheating (Chapter 9); and many, many more.

But what's the point of drawing connections between Bible stories and research on honesty? For me, there are two main points. The first is about specific, practical ideas for maintaining and improving honesty in society, and the second is about the general lessons we can draw from religion.

The specific, practical lessons we can draw from religion are that moral reminders, rules, habits, and starting over can help all of us stay on the straight and narrow for a bit longer. From this perspective, we can wonder whether the recent trend we see in parts of Europe and North America of people becoming more spiritual but less religious is a move in the right direction. I personally suspect that there are many advantages to spirituality, but from the perspective of decreasing dishonesty, what we need are specific and precise rules that are ideally connected to a higher meaning. For example, it would be delightful if everyone adopted the Golden Rule, but since the Golden Rule is a general rule without a specific action plan, it is unlikely to be sufficiently effective. Perhaps what we should aim for is increased spirituality that is accompanied by more specific rules that ground general principles in our daily lives. Whether we can create such rules and whether they will be effective is an important and interesting question.

Another way to think about the lessons of religion with regard to honesty has to do with timelines. There's the period *before* we have a chance to cheat, the period *during* which we have a chance to cheat, and the period *after* we have a chance to cheat. In which of these three periods is it best to apply the moral brakes?

In the legal, rational world of today, society takes a punitive approach with an eye toward dissuading bad behavior. Accordingly, we tend to focus exclusively on what can happen to us *after* we have had a chance to cheat. This is why the most standard approach to dealing with texting while driving, for example, is to make sure that people understand the high fine (which, in my state of North Carolina, is a hefty $500). Our legal system imposes harsh prison sentences for drug dealing and other crimes; some countries and states impose the death penalty for murder.

The basic logic behind this kind of approach is the SMORC, which we discussed in Chapter 1. We believe potential criminals will think rationally about the possibility of receiving a large punishment if they get caught (which would, of course, take place *after* the crime). We think they will weigh the costs and benefits of committing the crime before they do anything, and they will rationally decide that the benefit of the crime is not worth the potential costs.

Clearly, this approach to reducing crime isn't very effective. The threat of giving people tickets hasn't stopped people from texting and driving. The threat of stints in jail hasn't stopped burglaries. The threat of time in prison hasn't stopped violent crime.

In contrast, the general approach of religion is to deal directly with the period *before* we cheat and the period in which we have the *opportunity* to cheat. First, religion attempts to influence our mind-set *before* we are tempted, by creating moral education and—let's not forget—guilt. The basic understanding is that if we want to curb dishonesty, we need to think about education and calibrating the moral compass, rather than threatening punishment after the fact (which many religions are also pretty clear about). Second, religions attempt to influence our mind-sets *in the moment* of temptation by incorporating different moral reminders into our environment. Here, the basic idea is that once we have a moral compass, it's a good idea to keep it in good working order, with appropriate adjustments in real time, if we expect it to operate at full capacity.

SO WHERE DOES all this leave us? The good news is that we all have a moral compass. The bad news is that we can't just assume that our consciences will effortlessly and continuously protect us. As a society, we need to figure out how to instill a moral compass in our kids and how to maintain our own. Can we eradicate bad behavior? Most likely not, but I think we can certainly do better than we are doing now.

Thanks

I find writing about academic research to be fulfilling and stimulating, but the pleasure that I get day in and day out comes from working jointly with amazing researchers/friends—coming up with ideas, designing experiments, finding out what works and doesn't work, and figuring out what the results mean. The research described here is largely a product of my collaborators' ingenuity and efforts (see the following biographies of my outstanding colleagues), and I am grateful that we have been able to travel together in the landscape of dishonesty and together learn a bit about this important and fascinating topic.

In addition, I am also thankful to social scientists at large. The world of social science is an exciting place in which new ideas are constantly generated, data collected, and theories revised (some more than others). Every day I learn new things from my fellow researchers and am reminded of how much I don't know (for a partial list of references and additional readings, see the end of this book).

This is my third book, and by now one might expect that I would know what I am doing. But the reality is that I would not be able to do much without the help of many people. My deepest thanks go to Erin Allingham, who helped me write; Bronwyn Fryer and Julianne Wurm, who helped me see more clearly; Claire Wachtel, who conducted the process with grace and humor that is rare in editors; Elizabeth Perrella and Katherine Beitner, who managed to be my human substitutes for both Adderall and Xanax. And the team at Levine Greenberg Literary Agency, who were there to help in every possible way. Aline Grüneisen made many suggestions, some that were very insightful and others that made me smile. I am also grateful to Ania Jakubek, Sophia Cui, and Kacie Kinzer. Very special thanks also go to the person who functions as my external memory, hands, and alter ego: Megan Hogerty.

Finally, where would I be without my lovely wife, Sumi? It takes a very special person to be willing to share a life with me, and my hectic life and workaholism don't make it any easier. Sumi, I will move the boxes to the attic when I get home tonight. Actually, I will probably be late, so I will do it tomorrow. Well, you know what? I will definitely do it this weekend. I promise.

Loving, Dan

List of Collaborators

Aline Grüneisen

Aline joined my research team soon after I moved to Duke, and she has been a major force of energy and excitement ever since. I am not sure if this is part of her plan, but over time I have found myself depending on her to a larger and larger degree. Aline and I have been working together on a broad range of topics, and the unifying theme of all of them is that they are innovative and fun. Aline is currently the lab manager of the Center for Advanced Hindsight at Duke University, and I hope she will continue working with me for many more years.

Ayelet Gneezy

I met Ayelet many years ago at a picnic organized by mutual friends. I had a very positive first impression of her, and my appreciation of her has only increased with time. Ayelet is a wonderful person and a great friend, so it is a bit odd that the topics we decided to collaborate on were mistrust and revenge. Whatever initially drove us to explore these topics ended up being very useful, both academically and personally. Ayelet is currently a professor at the University of California, San Diego.

David Pizarro

David and I first met at an academic summer retreat at Stanford University. We shared a wall between our offices, and that was my first real introduction to rap music. A few weeks into it, I started enjoying the music, and David was kind enough to share his music collection with me (not sure how legal this was). Over the years I have gotten to spend a lot of time with David, and I always learn a lot, get energized, and wish I had more time with him. David is currently a professor at Cornell University.

Eynav Maharabani

I met Eynav in one of my visits to Israel. At the time she was a graduate student who just started working with Racheli Barkan. I was very impressed with her mix of intelligence, politeness, and assertiveness from the get-go, and it is the mix of these abilities that made her such a wonderful collaborator. Eynav is currently working at Abilities Solution, a unique company that focuses on employing people with disabilities for high-tech companies.

Francesca Gino

Francesca is a rare combination of kindness, caring, knowledge, creativity, and style. She also has endless energy and enthusiasm, and the number of projects she is involved with at any one time is generally what other people do in a lifetime. As an Italian, she is also one of the best people to share a meal and wine with. It was a deeply sad day for me when she decided to move from North Carolina to Boston. Francesca is currently a professor at Harvard University.

Janet Schwartz

I was lucky enough to tempt Janet to spend a few years with me at the Center for Advanced Hindsight. Janet is particularly interested in irrationalities related to health care (of which there are many), and together we have explored eating, dieting, advice, conflicts of interests, second opinions, and different approaches to getting people to behave as if they cared about their long-term health. Janet has a keen sense of observation about the world around her, and she is a fantastic storyteller, making fun of herself and everyone around her. Janet is currently a professor at Tulane University, but in spirit she is still at The Center.

Lisa Shu

Lisa is as bright as she is fun to be with. She has a sort of sixth sense for food, good research ideas, and fashion. These qualities make her not only a perfect collaborator but also a great shopping partner. In addition to studying ethical behavior she is interested in negotiation. And although I have never had the opportunity to personally negotiate with her, I have no doubt that if I did I would lose badly. Lisa is currently a PhD student at Harvard University.

Mary Frances Luce

Mary Frances was a PhD student at Duke a few years ahead of me and came back to Duke as a faculty member, also a few years ahead of me. Naturally this has made her a good source for advice over the years, and she has always been very supportive and helpful. A few years ago she moved to the dean's office, and both for my own sake and for the sake of the school, I hope that I am not going to continue following in her footsteps. Mary Frances is currently a professor at Duke University.

Maurice Schweitzer

Maurice finds almost anything around him interesting, and he approaches new projects with a big smile and great curiosity. For years now he has told me that he is a good squash player, and though I want to personally check how good he really is, I am also a bit worried that I will find out that he is much better than me. Maurice is also always a good source of wisdom on work, family, and life. Maurice is currently a professor at the University of Pennsylvania.

Max Bazerman

Max is insightful about pretty much any topic that comes up in research, politics, and personal life. And he always has something unexpected and interesting to say. After finding out that many of his students solve their own dilemmas and make decisions by asking themselves, "What would Max do?," I tried this approach a few times myself and can attest to its usefulness. Max is currently a professor at Harvard University.

Michael Norton

Mike is an interesting mix of brilliance, self-deprecation, and a sarcastic sense of humor. He has a unique perspective on life, and he finds almost any topic interesting. Mike is a great person to bounce ideas off of, and his feedback is always a mix of wacky, unexpected, insightful, and constructive. I often think about research projects as journeys, and with Mike I get to go on adventures that would be impossible with anyone else. Mike is currently a professor at Harvard University.

Nicole Mead

I first met Nicole when she was a graduate student at Florida State University. It was late, after a lecture I gave, and we ended up drinking a bit too much. I remember that I was very impressed with the ideas we were exchanging, but at some point I asked Nicole whether she thought that they were really good ideas or if it was the alcohol. Nicole assured me that it was not the alcohol, and I think she was mostly correct. Nicole has had many good ideas then and since, and she is currently a professor at Católica-Lisbon in Portugal.

Nina Mazar

Nina first came to MIT for a few days to get feedback on her research and ended up staying for five years. During this time we had oodles of fun working together and I came to rely on her greatly. Nina is impervious to obstacles, and her willingness to take on large challenges led us to carry out some particularly difficult experiments in rural India. For many years I hoped that she would never decide to leave, but, alas, the time came. She is currently a professor at the University of Toronto. In an alternate reality, Nina is a high-fashion designer in Milan.

On Amir

On joined MIT as a PhD student a year after I joined as a new professor and became "my" first student. In that capacity, he had a tremendous role in shaping what I expect from students and how I see the professor-student relationship. In addition to being exceptionally smart, On has an amazing set of skills, and what he does not know he is able to learn within a day or two. It is always exciting to work and spend time with him. On is currently a professor at the University of California, San Diego.

Racheli Barkan

Racheli (Rachel) and I became friends many years ago when we were both graduate students. Over the years we talked about starting various research projects together, but we really only got started when she came to spend a year at Duke. As it turned out, coffee is an important ingredient for translating ideas into action, and we had lots of fun during her visit and made a lot of progress on a wide range of projects. Racheli is incredibly knowledgeable, smart, and insightful, and I only wish we had more time together. Racheli is currently a professor at Ben-Gurion University of the Negev in Israel.

Roy Baumeister

Roy is a unique mixture of philosopher, musician, poet, and keen observer of human life. His interests span everything, and his perspective often looks puzzling to me at first, but then I realize the wisdom in it and end up thinking about his views for a long while—often adopting them. Roy is an ideal person to travel and explore with. He is currently a professor at Florida State University.

Scott McKenzie

Scott was an enthusiastic Duke undergraduate when he joined the Center for Advanced Hindsight. He was highly social and had a natural knack for getting people to do what he wanted them to do, including participating in our studies. When it was time for him to pick a topic for an independent research project, he picked cheating in golf, and through the process I learned a lot about that noble game. Scott is currently putting in the consulting world.

Shahar Ayal

I first met Shahar socially through common friends and then again when he was studying for his PhD under the supervision of another friend. So when he graduated, our personal and professional paths combined and he came to spend a few years at the Center for Advanced Hindsight as a postdoctoral fellow. Over those years we got to understand each other to a deeper level and think even more alike (mostly for the better). Shahar is a delight to be with and work with, and I am looking forward to many years of joint research. Shahar is currently a professor at the Interdisciplinary Institute in Israel.

Tom Gilovich

When I was a PhD student, I attended one of Tom's presentations, and I was amazed by the quality of his thought and creativity. Tom has a unique ability to ask important questions and find answers in interesting places. For example, he has shown that teams with black uniforms receive more penalties than their opponents; that basketball players don't really get a "hot hand"; and that NBA players miss more free throws when they don't think they deserve the penalty. I have always wanted to be a bit more like Tom. He is currently a professor at Cornell University.

Yoel Inbar

I first met Yoel when he was a student of Tom Gilovich and David Pizarro, and that is how we started working together. Yoel is the epitome of the modern hipster—equal parts cool and geek with a deep knowledge of indie rock bands (you probably haven't heard of them) and UNIX. One of Yoel's interests is disgust, and he is an expert in finding interesting ways to disgust people (fart spray, feces-shaped chocolate, odd foods, etc.). Yoel is currently a professor at Tilburg University in the Netherlands.

Zoë Chance

Zoë is a force of creativity and kindness. Talking to her is a bit like being in an amusement park—you know it is going to be exciting and interesting, but it is hard to anticipate which direction her comments will take. Together with her love of life and mankind, she is the ideal blend of researcher and friend. Zoë is currently a postdoctoral fellow at Yale University.

Notes

Introduction: Why Is Dishonesty So Interesting?

1. Ira Glass, "See No Evil," *This American Life*, National Public Radio, April 1, 2011.

Chapter 1. Testing the Simple Model of Rational Crime (SMORC)

1. "Las Vegas Cab Drivers Say They're Driven to Cheat," *Las Vegas Sun*, January 31, 2011, www.lasvegassun.com/news/2011/jan/31/driven-cheat/.

Chapter 3. Blinded by Our Own Motivations

1. A. Wazana, "Physicians and the Pharmaceutical Industry: Is a Gift Ever Just a Gift?" *Journal of the American Medical Association* (2000).
2. Duff Wilson, "Harvard Medical School in Ethics Quandary," *The New York Times*, March 2, 2009.

Chapter 5. Why Wearing Fakes Makes Us Cheat More

1. K. J. Winstein, "Inflated Credentials Surface in Executive Suite," *The Wall Street Journal*, November 13, 2008.

Chapter 6. Cheating Ourselves

1. Anne Morse, "Whistling Dixie," *The Weekly Standard* (blog), November 10, 2005.
2. Geoff Baker, "Mark McGwire Admits to Steroids Use: Hall of Fame Voting Becoming a Pain in the Exact Place He Used to Put the Needle,"

http://seattletimes.nwsource.com/html/marinersblog/2010767251_mark_mcgwire_admits_to_steroid.html.

Chapter 8. Cheating as an Infection: How We Can Catch the Dishonesty Germ

1. Steve Henn, "Oh, Waiter! Charge It to My PAC," *Marketplace*, July 21, 2008, and "PACs Put the Fun in Fundraising," *Marketplace*, July 22, 2008.
2. Steve Henn, "PACs Put the Fun in Fundraising," *Marketplace*, July 22, 2008.

Chapter 9. Collaborative Cheating

1. Dennis J. Devine, Laura D. Clayton, Jennifer L. Philips, Benjamin B. Dunford, and Sarah P. Melner, "Teams in Organizations, Prevalence, Characteristics, and Effectiveness," *Small Group Research* (1999).

   John Gordon, "Work Teams: How Far Have They Come?" *Training* (1992).

   Gerald E. Ledford, Jr., Edward E. Lawler III, and Susan A. Mohrman, "Reward Innovations in Fortune 1000 Companies," *Compensation & Benefits Review* (1995).

   Susan A. Mohrman, Susan G. Cohen, and Allan M. Mohrman, Jr., *Designing Team-Based Organizations: New Forms for Knowledge Work* (San Francisco: Jossey-Bass, 1995).

   Greg L. Stewart, Charles C. Manz, and Henry P. Sims, *Team Work and Group Dynamics* (New York: Wiley, 1999).
2. Bernard Nijstad, Wolfgang Stroebe, and Hein F. M. Lodewijkx, "The Illusion of Group Productivity: A Reduction of Failures Explanation," *European Journal of Social Psychology* (2006).
3. ADA Council on Scientific Affairs, "Direct and Indirect Restorative Materials," *The Journal of the American Dental Association* (2003).

Chapter 10. A Semioptimistic Ending: People Don't Cheat Enough!

1. *Montpelier* [Vermont] *Argus & Patriot*, March 6, 1873.

Bibliography and Additional Readings

Introduction: Why Is Dishonesty So Interesting?

Based on

Tim Harford, *The Logic of Life: The Rational Economics of an Irrational World* (New York: Random House, 2008).

Chapter 1. Testing the Simple Model of Rational Crime (SMORC)

Based on

Jerome K. Jerome, *Three Men in a Boat (to Say Nothing of the Dog)* (1889; reprint, New York: Tom Doherty Associates, 2001).

Jeff Kreisler, *Get Rich Cheating: The Crooked Path to Easy Street* (New York: HarperCollins, 2009).

Eynav Maharábani, "Honesty and Helping Behavior: Testing Situations Involving Temptation to Cheat a Blind Person," master's thesis, Ben-Gurion University of the Negev, Israel (2007).

Nina Mazar, On Amir, and Dan Ariely, "The Dishonesty of Honest People: A Theory of Self-concept Maintenance," *Journal of Marketing Research* (2008).

Nina Mazar and Dan Ariely, "Dishonesty in Everyday Life and Its Policy Implications," *Journal of Public Policy and Marketing* (2006).

Chapter 2. Fun with the Fudge Factor

Based on

Nina Mazar, On Amir, and Dan Ariely, "The Dishonesty of Honest People: A Theory of Self-concept Maintenance," *Journal of Marketing Research* (2008).

Lisa Shu, Nina Mazar, Francesca Gino, Max Bazerman, and Dan Ariely, "When to Sign on the Dotted Line? Signing First Makes Ethics Salient and Decreases Dishonest Self-Reports," working paper, Harvard Business School NOM Unit (2011).

Related readings

Jason Dana, Roberto A. Weber, and Jason Xi Kuang, "Exploiting Moral Wiggle Room: Behavior Inconsistent with a Preference for Fair Outcomes," *Economic Theory* (2007).

Christopher K. Hsee, "Elastic Justification: How Tempting but Task-Irrelevant Factors Influence Decisions," *Organizational Behavior and Human Decision Processes* (1995).

Christopher K. Hsee, "Elastic Justification: How Unjustifiable Factors Influence Judgments," *Organizational Behavior and Human Decision Processes* (1996).

Maurice Schweitzer and Chris Hsee, "Stretching the Truth: Elastic Justification and Motivated Communication of Uncertain Information," *The Journal of Risk and Uncertainty* (2002).

Chapter 2b. Golf

Related readings

Robert L. Goldstone and Calvin Chin, "Dishonesty in Self-report of Copies Made—Moral Relativity and the Copy Machine," *Basic and Applied Social Psychology* (1993).

Robert A. Wicklund, "The Influence of Self-awareness on Human Behavior," *American Scientist* (1979).

Chapter 3. Blinded by Our Own Motivations

Based on

Daylian M. Cain, George Loewenstein, and Don A. Moore, "The Dirt on Coming Clean: The Perverse Effects of Disclosing Conflicts of Interest," *Journal of Legal Studies* (2005).

Ann Harvey, Ulrich Kirk, George H. Denfield, and P. Read Montague, "Monetary Favors and Their Influence on Neural Responses and Revealed Preference," *The Journal of Neuroscience* (2010).

Related readings

James Bader and Daniel Shugars, "Agreement Among Dentists' Recommendations for Restorative Treatment," *Journal of Dental Research* (1993).

Max H. Bazerman and George Loewenstein, "Taking the Bias Out of Bean Counting," *Harvard Business Review* (2001).

Max H. Bazerman, George Loewenstein, and Don A. Moore, "Why Good Accountants Do Bad Audits: The Real Problem Isn't Conscious Corruption. It's Unconscious Bias," *Harvard Business Review* (2002).

Daylian M. Cain, George Loewenstein, and Don A. Moore, "When Sunlight Fails to Disinfect: Understanding the Perverse Effects of Disclosing Conflicts of Interest," *Journal of Consumer Research* (in press).

Carl Elliot, *White Coat, Black Hat: Adventures on the Dark Side of Medicine* (Boston: Beacon Press, 2010).

Chapter 4. Why We Blow It When We're Tired

Based on

Mike Adams, "The Dead Grandmother/Exam Syndrome and the Potential Downfall of American Society," *The Connecticut Review* (1990).

Shai Danziger, Jonathan Levav, and Liora Avnaim-Pesso, "Extraneous Factors in Judicial Decisions," *Proceedings of the National Academy of Sciences of the United States of America* (2011).

Nicole L. Mead, Roy F. Baumeister, Francesca Gino, Maurice E. Schweitzer, and Dan Ariely, "Too Tired to Tell the Truth: Self-Control Resource Depletion and Dishonesty," *Journal of Experimental Social Psychology* (2009).

Emre Ozdenoren, Stephen W. Salant, and Dan Silverman, "Willpower and the Optimal Control of Visceral Urges," *Journal of the European Economic Association* (2011).

Baba Shiv and Alexander Fedorikhin, "Heart and Mind in Conflict: The

Interplay of Affect and Cognition in Consumer Decision Making," *The Journal of Consumer Research* (1999).

Related readings

Roy F. Baumeister and John Tierney, *Willpower: Rediscovering the Greatest Human Strength* (New York: The Penguin Press, 2011).

Roy F. Baumeister, Kathleen D. Vohs, and Dianne M. Tice, "The Strength Model of Self-control," *Current Directions in Psychological Science* (2007).

Francesca Gino, Maurice E. Schweitzer, Nicole L. Mead, and Dan Ariely, "Unable to Resist Temptation: How Self-Control Depletion Promotes Unethical Behavior," *Organizational Behavior and Human Decision Processes* (2011).

C. Peter Herman and Janet Polivy, "A Boundary Model for the Regulation of Eating," *Research Publications—Association for Research in Nervous and Mental Disease* (1984).

Walter Mischel and Ozlem Ayduk, "Willpower in a Cognitive-Affective Processing System: The Dynamics of Delay of Gratification," in *Handbook of Self-regulation: Research, Theory, and Applications*, edited by Kathleen D. Vohs and Roy F. Baumeister (New York: Guilford, 2011).

Janet Polivy and C. Peter Herman, "Dieting and Binging, A Causal Analysis," *American Psychologist* (1985).

Chapter 5. Why Wearing Fakes Makes Us Cheat More

Based on

Francesca Gino, Michael I. Norton, and Dan Ariely, "The Counterfeit Self: The Deceptive Costs of Faking It," *Psychological Science* (2010).

Related readings

Dan Ariely and Michael L. Norton, "How Actions Create—Not Just Reveal—Preferences," *Trends in Cognitive Sciences* (2008).

Roy F. Baumeister, Kathleen D. Vohs, and Dianne M. Tice, "The Strength Model of Self-control," *Current Directions in Psychological Science* (2007).

C. Peter Herman and Deborah Mack, "Restrained and Unrestrained Eating," *Journal of Personality* (1975).

Chapter 6. Cheating Ourselves

Based on

Zoë Chance, Michael I. Norton, Francesca Gino, and Dan Ariely, "A Temporal View of the Costs and Benefits of Self-Deception," *Proceedings of the National Academy of Sciences* (2011).

Related readings

Ziva Kunda, "The Case for Motivated Reasoning," *Psychological Bulletin* (1990).

Danica Mijović-Prelec and Dražen Prelec, "Self-deception as Self-Signalling: A Model and Experimental Evidence," *Philosophical Transactions of the Royal Society* (2010).

Robert Trivers, "The Elements of a Scientific Theory of Self-Deception," *Annals of the New York Academy of Sciences* (2000).

Chapter 7. Creativity and Dishonesty: We Are All Storytellers

Based on

Edward J. Balleisen, "Suckers, Swindlers, and an Ambivalent State: A History of Business Fraud in America," manuscript.

Shane Frederick, "Cognitive Reflection and Decision Making," *Journal of Economic Perspectives* (2005).

Michael S. Gazzaniga, "Consciousness and the Cerebral Hemispheres," in *The Cognitive Neurosciences*, edited by Michael S. Gazzaniga (Cambridge, Mass.: MIT Press, 1995).

Francesca Gino and Dan Ariely, "The Dark Side of Creativity: Original Thinkers Can Be More Dishonest," *Journal of Personality and Social Psychology* (2011).

Ayelet Gneezy and Dan Ariely, "Don't Get Mad, Get Even: On Consumers' Revenge," working paper, Duke University (2010).

Richard Nisbett and Timothy DeCamp Wilson, "Telling More Than We Can Know: Verbal Reports on Mental Processes," *Psychological Review* (1977).

Yaling Yang, Adrian Raine, Todd Lencz, Susan Bihrle, Lori Lacasse, and Patrick Colletti, "Prefrontal White Matter in Pathological Liars," *The British Journal of Psychiatry* (2005).

Related readings

Jesse Preston and Daniel M. Wegner, "The Eureka Error: Inadvertent Plagiarism by Misattributions of Effort," *Journal of Personality and Social Psychology* (2007).

Chapter 8. Cheating as an Infection: How We Catch the Dishonesty Germ

Based on

Nicholas A. Christakis and James H. Fowler, *Connected: The Surprising Power of Our Social Networks and How They Shape Our Lives* (New York: Little, Brown, 2009).

Robert B. Cialdini, *Influence: The Psychology of Persuasion* (New York: William Morrow, 1993).

Francesca Gino, Shahar Ayal, and Dan Ariely, "Contagion and Differentiation in Unethical Behavior: The Effect of One Bad Apple on the Barrel," *Psychological Science* (2009).

George L. Kelling and James Q. Wilson, "Broken Windows: The Police and Neighborhood Safety," *The Atlantic* (March 1982).

Nina Mazar, Kristina Shampanier, and Dan Ariely, "Probabilistic Price Promotions—When Retailing and Las Vegas Meet," working paper, Rotman School of Management, University of Toronto (2011).

Related readings

Ido Erev, Paul Ingram, Ornit Raz, and Dror Shany, "Continuous Punishment and the Potential of Gentle Rule Enforcement," *Behavioural Processes* (2010).

Chapter 9. Collaborative Cheating: Why Two Heads Aren't Necessarily Better than One

Based on

Melissa Bateson, Daniel Nettle, and Gilbert Roberts, "Cues of Being Watched Enhance Cooperation in a Real-World Setting," *Biology Letters* (2006).

Francesca Gino, Shahar Ayal, and Dan Ariely, "Out of Sight, Ethically Fine? The Effects of Collaborative Work on Individuals' Dishonesty," working paper (2009).

Janet Schwartz, Mary Frances Luce, and Dan Ariely, "Are Consumers

Too Trusting? The Effects of Relationships with Expert Advisers," *Journal of Marketing Research* (2011).

Related readings

Francesca Gino and Lamar Pierce, "Dishonesty in the Name of Equity," *Psychological Science* (2009).

Uri Gneezy, "Deception: The Role of Consequences," *American Economic Review* (2005).

Nina Mazar and Pankaj Aggarwal, "Greasing the Palm: Can Collectivism Promote Bribery?" *Psychological Science* (2011).

Scott S. Wiltermuth, "Cheating More When the Spoils Are Split," *Organizational Behavior and Human Decision Processes* (2011).

Chapter 10. A Semioptimistic Ending: People Don't Cheat Enough!

Based on

Rachel Barkan and Dan Ariely, "Worse and Worst: Daily Dishonesty of Business-men and Politicians," working paper, Ben-Gurion University of the Negev, Israel (2008).

Yoel Inbar, David Pizarro, Thomas Gilovich, and Dan Ariely, "Moral Masochism: Guilt Causes Physical Self-punishment," working paper (2009).

Azim Shariff and Ara Norenzayan, "Mean Gods Make Good People: Different Views of God Predict Cheating Behavior," *International Journal for the Psychology of Religion* (2011).

Related readings

Keri L. Kettle and Gerald Häubl, "The Signature Effect: How Signing One's Name Influences Consumption-Related Behavior by Priming Self-Identity," *Journal of Consumer Research* (2011).

Deepak Malhotra, "(When) Are Religious People Nicer? Religious Salience and the 'Sunday Effect' on Pro-Social Behavior," *Judgment and Decision Making* (2010).

Index

About the Author

Dan Ariely is the James B. Duke Professor of Psychology and Behavioral Economics at Duke University, with appointments at the Fuqua School of Business, the Center for Cognitive Neuroscience, the Department of Economics, and the School of Medicine. Dan earned one PhD in cognitive psychology and another PhD in business administration. He is the founder and director of the Center for Advanced Hindsight. His work has been featured in many outlets, including *The New York Times*, *The Wall Street Journal*, *The Washington Post*, *The Boston Globe*, and others. He lives in Durham, North Carolina, with his wife, Sumi, and their two creative children, Amit and Neta.

www.danariely.com

BOOKS BY DAN ARIELY

THE (HONEST) TRUTH ABOUT DISHONESTY

How We Lie to Everyone—Especially Ourselves

Available in Paperback and eBook

"A lively tour through the impulses that cause many of us to cheat, the book offers especially keen insights into the ways in which we cut corners while still thinking of ourselves as moral people."
—*Time*

THE UPSIDE OF IRRATIONALITY

The Unexpected Benefits of Defying Logic

Available in Paperback, eBook, and Audio CD

"*The Upside of Irrationality* is an eye-opening, insightful look at human behavior, proving that defying logic is part of what makes us human."
—*Boston Globe*

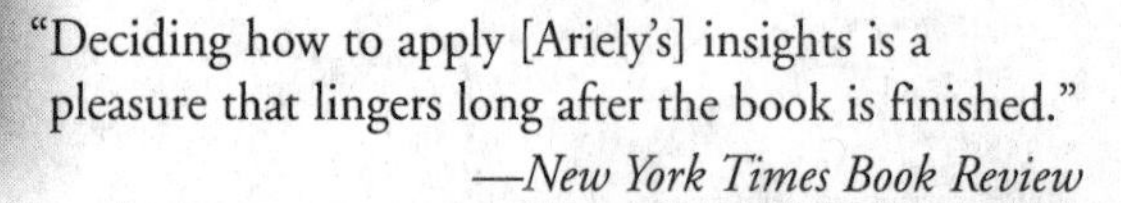

"Deciding how to apply [Ariely's] insights is a pleasure that lingers long after the book is finished."
—*New York Times Book Review*

PREDICTABLY IRRATIONAL

The Hidden Forces That Shape Our Decisions

Available in Paperback, eBook, and Audio CD

"A fascinating romp through the science of decision-making that unmasks the ways that emotions, social norms, expectations, and context lead us astray." —*Time*

"Surprisingly entertaining . . . easy to read . . . Ariely's book makes economics and the strange happenings of the human mind fun."
—*USA Today*